From Complex Sentences to a Formal Semantic Representation using Syntactic Text Simplification and Open Information Extraction

Christina Niklaus

From Complex Sentences to a Formal Semantic Representation using Syntactic Text Simplification and Open Information Extraction

 Springer Vieweg

Christina Niklaus
Rehetobel, Switzerland

This text is a reprint of a dissertation submitted in 2021 to the Faculty of Computer Science and Mathematics at the University of Passau in partial fulfillment of the requirements for the degree of Doctor of Natural Sciences.

ISBN 978-3-658-38696-2   ISBN 978-3-658-38697-9 (eBook)
https://doi.org/10.1007/978-3-658-38697-9

Responsible Editor: Stefanie Eggert
This Springer Vieweg imprint is published by the registered company Springer Fachmedien Wiesbaden GmbH, part of Springer Nature.
The registered company address is: Abraham-Lincoln-Str. 46, 65189 Wiesbaden, Germany

Acknowledgements

This thesis is the result of a long journey. There are many who helped me along the way. I want to take a moment to thank them.

First, I would like to express my sincere gratitude to my supervisor Prof. Siegfried Handschuh for the continuous support of my work, for his patience, motivation, and belief in me. I am deeply grateful for your assistance at every stage of this project. Thank you for your invaluable advice, insightful comments and suggestions that have encouraged me throughout my research and writing of this thesis. Without your guidance and constant feedback, this dissertation would have never been accomplished. *Danke!*

I would also like to thank Dr. André Freitas, whose guidance, support, and encouragement have been invaluable throughout my studies. You provided me with the tools that I needed to choose the right direction and successfully complete my dissertation. Your insightful feedback pushed me to sharpen my thinking and brought my work to a higher level. You went above and beyond to help me reach my goal. I could not have imagined a better advisor and mentor for my Ph.D. study. *Obrigada!*

Moreover, I would like to acknowledge my fellow labmates from the Data Science and Natural Language Processing group at the University of St.Gallen, as well as my former colleagues back at the University of Passau. Thank you for all the stimulating discussions that shaped and improved some of the ideas of this thesis. In particular, I would like to thank my office mates Matthias, Thomas, Fabian, and Reto for their support and the happy distractions in stressful times. You were fun to hang out with and made this place a productive and joyful working environment!

To my parents: you set me off on the road to this Ph.D. a long time ago by always encouraging me to follow my goals—even if it meant moving further and

v

further away. Thank you for always being there for me. You gave me the best shelter I could have imagined during the stressful time of the Covid-19 pandemic when I was writing up this thesis. I will now clear all the papers off the kitchen table as I promised! I am forever grateful for your patience and understanding. *Danke, Mama & Papa!*

Finally, to my husband, Andreas: you have supported me all the way to this dissertation, patiently enduring my stresses and moans for the past five years that I was working on this thesis. Thank you for putting up with me being sat in the office for hours on end, including many weekends, to meet the next deadline. Your love and understanding helped me through difficult times. Without you believing in me, I never would have made it. Thank you for your tremendous patience and encouragement. You are amazing! It is now time to celebrate; you earned this degree right along with me. *Ich liebe Dich!*

Abstract

Sentences that present a complex linguistic structure act as a major stumbling block for Natural Language Processing (NLP) applications whose predictive quality deteriorates with sentence length and complexity. The task of Text Simplification (TS) may remedy this situation. It aims to modify sentences in order to make them easier to process, using a set of rewriting operations, such as reordering, deletion or splitting. These transformations are executed with the objective of converting the input into a simplified output, while preserving its main idea and keeping it grammatically sound.

State-of-the-art syntactic TS approaches suffer from two major drawbacks: first, they follow a very conservative approach in that they tend to retain the input rather than transforming it, and second, they ignore the cohesive nature of texts, where context spread across clauses or sentences is needed to infer the true meaning of a statement. To address these problems, we present a discourse-aware TS framework that is able to split and rephrase complex English sentences within the semantic context in which they occur. By generating a fine-grained output with a simple canonical structure that is easy to analyze by downstream applications, we tackle the first issue. For this purpose, we decompose a source sentence into smaller units by using a linguistically grounded transformation stage. The result is a set of self-contained propositions, with each of them presenting a minimal semantic unit. To address the second concern, we suggest not only to split the input into isolated sentences, but to also incorporate the semantic context in the form of hierarchical structures and semantic relationships between the split propositions. In that way, we generate a semantic hierarchy of minimal propositions that benefits downstream Open Information Extraction (IE) tasks. To function well, the TS approach that we propose requires syntactically well-formed input sentences. It targets general-purpose texts in English, such as newswire or Wikipedia articles, which commonly contain a high proportion of complex assertions.

In a second step, we present a method that allows state-of-the-art Open IE systems to leverage the semantic hierarchy of simplified sentences created by our discourse-aware TS approach in constructing a lightweight semantic representation of complex assertions in the form of semantically typed predicate-argument structures. In that way, important contextual information of the extracted relations is preserved that allows for a proper interpretation of the output. Thus, we address the problem of extracting incomplete, uninformative or incoherent relational tuples that is commonly to be observed in existing Open IE approaches. Moreover, assuming that shorter sentences with a more regular structure are easier to process, the extraction of relational tuples is facilitated, leading to a higher coverage and accuracy of the extracted relations when operating on the simplified sentences. Aside from taking advantage of the semantic hierarchy of minimal propositions in existing Open IE approaches, we also develop an Open IE reference system, Graphene. It implements a relation extraction pattern upon the simplified sentences.

The framework we propose is evaluated within our reference TS implementation DISSIM. In a comparative analysis, we demonstrate that our approach outperforms the state of the art in structural TS both in an automatic and a manual analysis. It obtains the highest score on three simplification datasets from two different domains with regard to SAMSA (0.67, 0.57, 0.54), a recently proposed metric targeted at automatically measuring the syntactic complexity of sentences which highly correlates with human judgments on structural simplicity and grammaticality. These findings are supported by the ratings from the human evaluation, which indicate that our baseline implementation DISSIM returns fine-grained simplified sentences that achieve a high level of syntactic correctness and largely preserve the meaning of the input. Furthermore, a comparative analysis with the annotations contained in the RST Discourse Treebank (RST-DT) reveals that we are able to capture the contextual hierarchy between the split sentences with a precision of approximately 90% and reach an average precision of almost 70% for the classification of the rhetorical relations that hold between them. Finally, an extrinsic evaluation shows that when applying our TS framework as a pre-processing step, the performance of state-of-the-art Open IE systems can be improved by up to 32% in precision and 30% in recall of the extracted relational tuples.

Accordingly, we can conclude that our proposed discourse-aware TS approach succeeds in transforming sentences that present a complex linguistic structure into a sequence of simplified sentences that are to a large extent grammatically correct, represent atomic semantic units and preserve the meaning of the input. Moreover, the evaluation provides sufficient evidence that our framework is able to establish a semantic hierarchy between the split sentences, generating a fine-grained representation of complex assertions in the form of hierarchically ordered and semantically interconnected propositions. Finally, we demonstrate that state-of-the-art Open IE systems benefit from using our TS approach as a pre-processing step by increasing both the accuracy and coverage of the extracted relational tuples for the majority of the Open IE approaches under consideration. In addition, we outline that the semantic hierarchy of simplified sentences can be leveraged to enrich the output of existing Open IE systems with additional meta information, thus transforming the shallow semantic representation of state-of-the-art approaches into a canonical context-preserving representation of relational tuples.

Contents

Acronyms

AMR	Abstract Meaning Representation
AUC	area under the PR curve
CRF	Conditional Random Field
CSD	contextual sentence decomposition
Decomp	Universal Decompositional Semantics
DRS	Discourse Representation Structure
DRT	Discourse Representation Theory
DSS	Direct Semantic Splitting
EDU	elementary discourse unit
EW	English Wikipedia
FKGL	Flesch-Kincaid Grade Level
IE	Information Extraction
ILP	Integer Linear Programming
KBP	Knowledge Base Population
LSTM	Long Short-Term Memory
ML	Machine Learning
MR	meaning representation
MT	Machine Translation
MTL	Multi-Task Learning
NER	Named Entity Recognition
NL	natural language
NLP	Natural Language Processing
NMT	neural machine translation
NSE	Neural Semantic Encoder
NTS	Neural Text Simplification
PBMT	phrase-based machine translation

PDTB	Penn Discourse TreeBank
PR	precision-recall
PWKP	Parallel Wikipedia Simplification
QA	Question Answering
QAMR	Question-Answer Meaning Representation
QG	Quasi-synchronous Grammar
RDF	Resource Description Framework
RL	Reinforcement Learning
RNN	Recurrent Neural Network
RST	Rhetorical Structure Theory
RST-DT	RST Discourse Treebank
SBMT	syntax-based machine translation
SDRT	Segmented Discourse Representation Theory
Seq2Seq	sequence-to-sequence
SEW	Simple English Wikipedia
SL	Sequence Labeling
SMT	statistical machine translation
SRL	Semantic Role Labeling
STSG	Synchronous Tree Substitution Grammar
TS	Text Simplification
TUPA	Transition-based UCCA Parser
UCCA	Universal Conceptual Cognitive Annotation

List of Figures

List of Tables

Part I
Background

"The ability to simplify means to eliminate the unnecessary so that the necessary may speak."
Hans Hofmann (1880–1966)

Introduction

1

Modern systems that deal with inference in texts need automatized methods to extract meaning representations (MRs) from texts at scale. Open IE is a prominent way of extracting all potential relations from a given text in a comprehensive manner. However, sentences that present a complex linguistic structure can be *difficult to analyze by NLP applications* (Mitkov and Saggion, 2018). Identifying grammatical complexities in a sentence and transforming them into simpler structures, using a set of text-to-text rewriting operations, is the goal of syntactic TS. The major types of operations that are used to perform this rewriting step are *deletion* and *sentence splitting*. The goal of the deletion task is to remove parts of a sentence that contain only peripheral information in order to produce an output proposition that is more succinct and contains only the key information of the input. The splitting operation, on the contrary, aims at preserving the complete information of the input. Therefore, it divides a sentence into several shorter components, with each of them presenting a simpler and more regular structure that is easier to process by both humans and machines and may support a faster generalization in Machine Learning (ML) tasks.

Syntactic TS with a focus on the task of sentence splitting has been attracting growing interest in the NLP community within the past few years. It has been shown that sentence splitting may benefit both NLP and societal applications. One line of work encompasses syntactic TS approaches that serve as **assistive technology for specific target reader populations** (Ferrés et al., 2016; Siddharthan and Mandya, 2014; Saggion et al., 2015). The main goal of such approaches is to improve the readability of a text, i.e. to enhance the ease with which it can be understood. Thus, it aims to make information easier to comprehend for people with reading difficulties that may arise from, for example, aphasia (Carroll et al., 1999), autism (Evans et al., 2014) or deafness (Inui et al., 2003), as well as for those with reduced reading

© The Author(s), under exclusive license to Springer Fachmedien Wiesbaden GmbH, part of Springer Nature 2022
C. Niklaus, *From Complex Sentences to a Formal Semantic Representation using Syntactic Text Simplification and Open Information Extraction*,
https://doi.org/10.1007/978-3-658-38697-9_1

levels, including children (De Belder and Moens, 2010) and non-native speakers of the respective language (Angrosh et al., 2014). In that way, information is made available to a broader audience. Apart from substituting a word or phrase that is hard to understand with a more comprehensible synonym (*lexical simplification*), the most effective operations to improve the reading comprehension for language-impaired humans are splitting long sentences, making discourse relations explicit, avoiding pre-posed adverbial clauses and presenting information in cause-effect order (Siddharthan, 2014).

The second line of work aims at **generating an intermediate representation that is easier to analyze by downstream applications whose predictive quality deteriorates with sentence length and structural complexity**, posing problems due to their potential high level of ambiguity. Prior work has established that shortening sentences by dropping constituents or splitting components, and thus operating on smaller units of text, facilitates and improves the performance of a variety of applications, including Machine Translation (MT) (Štajner and Popovic, 2016; Štajner and Popović, 2018), Open IE (Cetto et al., 2018b; Saha and Mausam, 2018), Text Summarization (Siddharthan et al., 2004; Bouayad-Agha et al., 2009), Relation Extraction (Miwa et al., 2010), Semantic Role Labeling (SRL) (Vickrey and Koller, 2008), Question Generation (Heilman and Smith, 2010; Bernhard et al., 2012), Sentence Fusion (Filippova and Strube, 2008) and Parsing (Chandrasekar et al., 1996).

Many different methods for addressing the task of TS have been proposed so far. As noted in Štajner and Glavaš (2017), data-driven approaches outperform rule-based systems in the area of lexical simplification, which aims at replacing a difficult word or phrase with a simpler synonym (Glavaš and Štajner, 2015; Paetzold and Specia, 2016; Nisioi et al., 2017; Zhang and Lapata, 2017). In contrast, the state-of-the-art structural simplification approaches are rule-based (Siddharthan and Mandya, 2014; Ferrés et al., 2016; Saggion et al., 2015), providing more grammatical output and covering a wider range of syntactic transformation operations, however, at the cost of being very conservative, often to the extent of not making any changes at all.

1.1   Semantic Hierarchy of Minimal Propositions through Discourse-Aware Sentence Splitting

In order to overcome the conservatism exhibited by state-of-the-art syntactic TS approaches, i.e. their tendency to retain the input sentence rather than transforming it, we propose a novel sentence splitting approach that **breaks down a complex**

English sentence into a set of minimal propositions, i.e. a sequence of grammatically sound, self-contained utterances, with each of them presenting a minimal semantic unit that cannot be further decomposed into meaningful propositions (Bast and Haussmann, 2013). For an example, see Table 1.1. Thus, we augment the Split-and-Rephrase task that was proposed in Narayan et al. (2017) by the notion of *minimality* (see Figure 1.1, subtask 1).

Table 1.1 Result of the sentence splitting subtask on an example sentence. A complex source sentence is decomposed into a set of minimal semantic units

Complex source	A fluoroscopic study which is known as an upper gastrointestinal series is typically the next step in management, although if volvulus is suspected, caution with non water soluble contrast is mandatory as the usage of barium can impede surgical revision and lead to increased post operative complications.
Minimal propositions	A fluoroscopic study is typically the next step in management. This fluoroscopic study is known as an upper gastrointestinal series. Volvulus is suspected. Caution with non water soluble contrast is mandatory. The usage of barium can impede surgical revision. The usage of barium can lead to increased post operative complications.

In the development of our approach, we followed a principled and systematic procedure, with the goal of eliciting a universal set of transformation rules for converting complex source sentences into a set of minimal propositions. The patterns were heuristically determined in a rule-engineering process that was carried out on the basis of an in-depth study of the literature on syntactic sentence simplification (Siddharthan, 2006, 2014, 2002; Siddharthan and Mandya, 2014; Evans, 2011; Evans and Orăsan, 2019; Heilman and Smith, 2010; Vickrey and Koller, 2008; Shardlow, 2014; Mitkov and Saggion, 2018; Mallinson and Lapata, 2019; Brouwers et al., 2014; Suter et al., 2016; Ferrés et al., 2016; Chandrasekar et al., 1996). We performed a thorough linguistic analysis of the syntactic phenomena that need to be tackled in the sentence splitting task and materialized our findings into a small set of 35 hand-crafted transformation rules that decompose each source sentence into a sequence of atomic semantic units. In that way, we generate self-contained propositions that present a simpler and grammatically sound structure. Nested structures are removed and related pieces of information are brought closer together, resulting in a fine-grained representation of the source sentence that is easier to process and thus leverages downstream Open IE applications, improving their performance in terms of accuracy and coverage.

SUBTASK 1: SPLIT INTO MINIMAL PROPOSITIONS

Input: A complex sentence C.

Problem: Produce a sequence of simple sentences $T_1, \ldots, T_n$, $n >= 2$, such that

1. each simple sentence T_i is grammtically sound. (**syntactic correctness**)

2. each simple sentence T_i presents a minimal semantic unit, i.e. it cannot be further decomposed into meaningful propositions. (**minimality**)

3. the output sentences $T_1, \ldots, T_n$ convey all and only the information in C. (**semantic correctness**)

SUBTASK 2: ESTABLISH A SEMANTIC HIERARCHY

Input: A pair of structurally simplified sentences T_i and T_j.

Problem: Establish a *contextual hierarchy* between the split sentences and capture their *semantic relationship*.

Figure 1.1 Problem description. Two subtasks are carried out to transform sentences that present a complex linguistic structure into a semantic hierarchy of minimal propositions. First, syntactically complex sentences are transformed into smaller units with a simpler structure, using clausal and phrasal disembedding mechanisms. Next, a hierarchical representation is generated between those units, capturing their semantic context

However, any sound and coherent text is not simply a loose arrangement of self-contained units, but rather a logical structure of utterances that are semantically connected (Siddharthan, 2014). Consequently, when carrying out syntactic simplification operations without considering discourse implications, the rewriting may easily result in a disconnected sequence of simplified sentences, making the text harder to interpret (see the output of the example given in Table 1.1). Still, the vast majority of existing structural TS approaches do not take into account discourse level effects that arise from applying syntactic transformations. I.e., they split the input into simplified sentences without considering and preserving their semantic context. Accordingly, the resulting simplified text is prone to lack coherence, consisting of a set of semantically unrelated utterances that miss important contextual information. Therefore, the interpretability of the output for downstream Open IE tasks may be hampered. Thus, in order to **preserve the coherence structure** and, hence, the

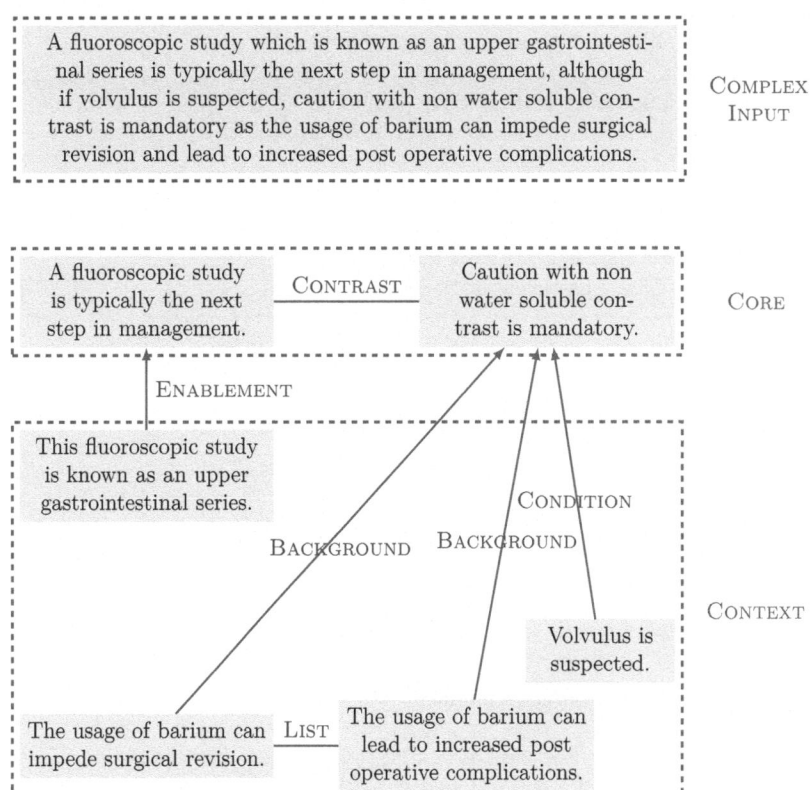

Figure 1.2 Example of the output that is generated by our proposed discourse-aware TS approach. A complex input sentence (see Table 1.1) is transformed into a semantic hierarchy of simplified sentences in the form of minimal, self-contained propositions that are linked via rhetorical relations. The output presents a regular, fine-grained structure that is easy to process, while still preserving the coherence and, hence, the interpretability of the output

interpretability of the output, we propose a discourse-aware TS approach based on Rhetorical Structure Theory (RST) (Mann and Thompson, 1988) that establishes a semantic hierarchy on the simplified sentences. For this purpose, we first set up a contextual hierarchy between the split components, distinguishing *core* sentences that contain the key information of the input from *contextual* sentences that disclose less relevant, supplementary information. Then, the semantic relationship that holds between the decomposed spans is identified and classified. Examples of the result-

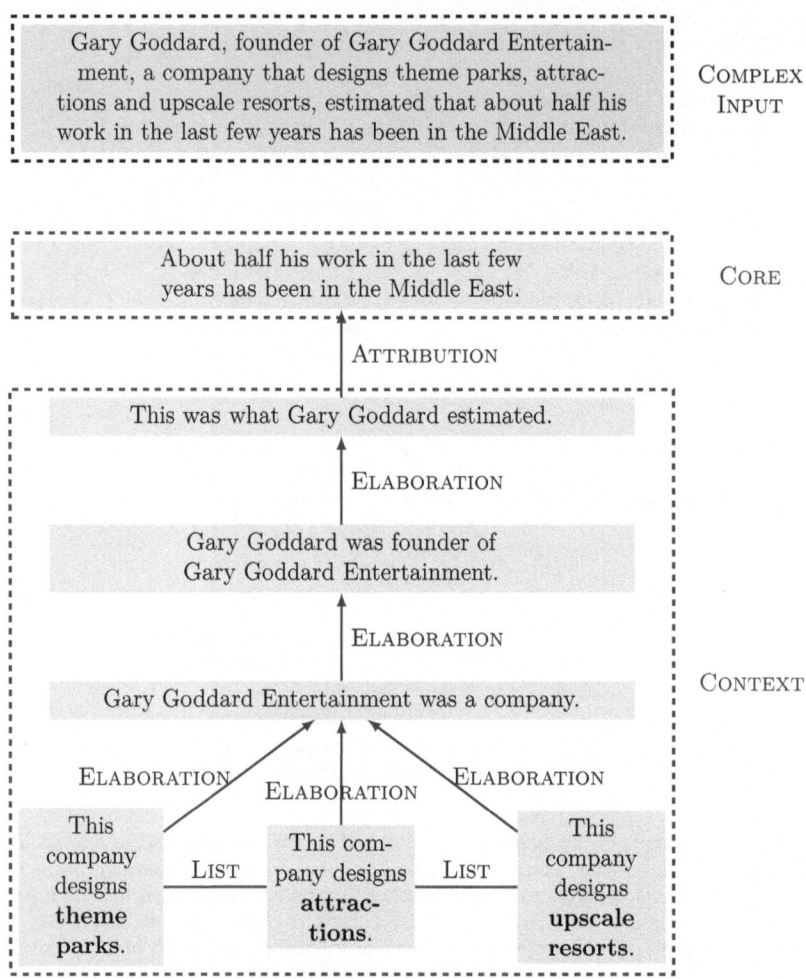

Figure 1.3 Context-preserving representation of the simplified sentences in the form of a semantic hierarchy of minimal propositions

ing context-preserving output are displayed in Figures 1.2 and 1.3. They represent two ideal-typical examples of a set of hierarchically ordered and semantically interconnected atomic propositions, illustrating different types of semantic relationships between the split utterances. In the following, the source sentence from Figure 1.2 will serve as our running example.

Using the discourse-aware TS approach described above, an intermediate representation of complex sentences in the form of a semantic hierarchy of minimal propositions is generated. In a second step, we leverage it for facilitating and improving the performance of Open IE applications, thus transforming complex sentences into a formal semantic representation.

1.2   Formal Semantic Representation through Open Information Extraction

After transforming sentences that present a complex linguistic structure into a semantic hierarchy of minimal propositions, a sentence-level MR is constructed on top of it. While tasks such as Open IE have been evolving along the years to capture relations beyond simple subject-predicate-object triples, a closer look into the extraction of the linkage between clauses and phrases within a complex sentence has been neglected. This work aims to address this gap. It is targeted on capturing complementary linguistic phenomena, including intra-sentential discourse relations, predicate-argument structures, as well as clausal and phrasal linkage patterns. By taking advantage of the split sentences instead of dealing directly with the raw source sentences, the complexity of this task will be reduced, facilitating the process of extracting a formal semantic representation from unstructured text.

More precisely, we leverage the semantic hierarchy of simplified sentences from the previous stage, where

- syntactically complex sentences were transformed into *smaller units with a simpler structure*, using clausal and phrasal disembedding mechanisms; and
- a hierarchical representation was generated between those units by applying a recursive top-down approach, *capturing both their semantic context and relations to other units* in the form of rhetorical relations.

We demonstrate that after this transformation process, relational tuples in terms of predicate-argument structures can be extracted more easily from the simplified sentences. In fact, as revealed in a comparative analysis with current Open IE systems, a single relation extraction rule is sufficient to achieve state-of-the-art results on a large benchmark corpus in terms of precision and recall.

In addition, our Open IE framework is able to **extract relations within the semantic context** in which they occur, enriching its output with important contextual information that is needed for a proper interpretation of complex assertions. We thus introduce an innovative lightweight semantic representation for relational tuples.

1.3   Problems in Syntactic Text Simplification and Open Information Extraction

As outlined in the sections above, six main problems can be identified in state-of-the-art syntactic TS and current Open IE approaches. Regarding the former, the following issues were recognized:

(1) **State-of-the-art TS approaches largely ignore the task of sentence splitting.** Traditionally, syntactic TS has been reduced to four rewriting operations: *substituting* complex words or phrases with simpler synonyms, *deleting* or *reordering* sentence components, and *splitting* a complex sentence into several simpler ones. However, in a recent survey, Alva-Manchego et al. (2020) showed that most research in TS has focused on studying replacement, reordering and deletion transformations, while the splitting operation is hardly investigated. Only 35% of the 20 TS approaches under consideration explicitly model sentence splitting, while as many as 95% of them cover substitutions, 75% reorderings and 70% deletions. Moreover, they observed that TS corpora, too, show a tendency towards lexical simplification and compression operations. For instance, the average number of simple sentences per complex source is 1.06 in the Parallel Wikipedia Simplification (PWKP) corpus (Zhu et al., 2010), which is one of the most widely used datasets in TS research (for details, see Section 2.1.2.1). In fact, only 6.1% of the complex sentences in PWKP are split into two or more simplified counterparts (Narayan and Gardent, 2014). To address this gap, Narayan et al. (2017) recently proposed the task of *Split-and-Rephrase*, whose goal is to split syntactically complex sentences into several shorter ones, while making the rephrasings that are necessary to preserve the meaning of the input, as well as the grammaticality of the simplified sentences. In this thesis, we aim to further explore this research direction, augmenting the Split-and-Rephrase task by the notion of *minimality*. For this purpose, we propose a new sentence splitting approach that breaks down a complex source sentence into a set of minimal propositions with a simpler and more regular structure. In addition, we compile a large-scale sentence splitting corpus of parallel complex source and simplified target sentences, which can be used for developing data-driven sentence splitting approaches that attempt to learn how to decompose complex sentences into a fine-grained representation of short sentences with a simplified syntax.

(2) **In the field of TS, there is little research on dealing with discourse-related issues caused by the rewriting transformations (Alva-Manchego et al., 2020).** The vast majority of existing structural TS approaches do not take into

account discourse-level aspects. Thus, they ignore the cohesive nature of texts, where context is propagated across sentences and clauses. Consequently, they are prone to producing a set of incoherent utterances that lack important contextual information that is needed to infer the true meaning of a statement. As opposed to those approaches, we propose a discourse-aware sentence splitting technique that aims to preserve the semantic relationship between the decomposed spans, thus maintaining their interpretability for downstream Open IE applications.

(3) **Current syntactic TS approaches follow a very conservative approach in that they tend to retain the input rather than transforming it.** When receiving a complex source sentence as input, state-of-the-art structural TS systems often do not perform any changes at all (Štajner and Glavaš, 2017). This poses a challenge to downstream Open IE applications, which regularly experience problems when having to deal with complex sentences that mix multiple, potentially semantically unrelated propositions (Bast and Haussmann, 2013). With our proposed TS approach, we aim to overcome this conservatism. Its objective is to split a complex input sentence into minimal propositions, with each of them presenting a separate fact, thus supporting machine processing in downstream Open IE tasks.

```
He nominated Sonia Sotomayor on May 26, 2009 to replace David Souter; she
was confirmed on August 6, 2009, becoming the first Supreme Court Justice
of Hispanic descent.

Ollie:
(1) she;                was confirmed on;               August 6, 2009
(2) He;                 nominated Sonia Sotomayor on;    May 26
(3) He;                 nominated Sonia Sotomayor;       2009
(4) He;                 nominated 2009 on;               May 26
(5) Sonia Sotomayor;    be nominated 2009 on;            May 26
(6) He;                 nominated 2009;                  Sonia Sotomayor
(7) 2009;               be nominated Sonia Sotomayor on; May 26


ClausIE:
(8) He;    nominated;    Sonia Sotomayor on May 26 2009 to replace
David Souter
(9) she;   was confirmed;  on August 6 2009 becoming the first Supreme
Court Justice of Hispanic descent
(10) she;  was confirmed;  becoming the first Supreme Court Justice of
Hispanic descent
```

Figure 1.4 Output generated by state-of-the-art Open IE systems

Regarding the task of Open IE, particularly long and syntactically complex sentences pose a challenge for current approaches. By analyzing the output of such systems, we observed three common shortcomings:

(4) **Relations often span over long nested structures or are presented in a non-canonical form that cannot be easily captured by a small set of extraction patterns. Therefore, such relations are commonly missed by state-of-the-art approaches.** For an example, consider the sentence from Figure 1.4, which asserts that ⟨*Sonia Sotomayor*; *became*; *the first Supreme Court Justice of Hispanic descent*⟩. This information is encoded in a complex participial construction that is omitted by both reference Open IE systems, OLLIE (Mausam et al., 2012) and ClausIE (Del Corro and Gemulla, 2013). By taking advantage of the split sentences instead of dealing directly with the raw source sentences, we aim to improve the coverage and accuracy of the extracted relational tuples.

(5) **Current Open IE systems tend to extract relational tuples with long argument phrases that can be further decomposed into meaningful propositions, with each of them representing a separate fact.** Overly specific constituents that mix multiple, potentially semantically unrelated propositions are difficult to handle for downstream applications, such as Question Answering (QA) or textual entailment tasks. Instead, such approaches benefit from extractions that are as compact as possible (Bast and Haussmann, 2013). This phenomenon can be witnessed particularly well in the extractions generated by ClausIE, whose argument phrases frequently combine several semantically independent statements. For instance, the argument in relational tuple (8) in Figure 1.4 contains three unrelated facts, namely a direct object ⟨*Sonia Sotomayor*⟩, a temporal expression ⟨*on May 26 2009*⟩ and a phrasal description ⟨*to replace David Souter*⟩ specifying the purpose of the assertion on which it depends. Here again, we leverage the minimal propositions resulting from the TS process, which allows us to operate on smaller units with a simplified structure, thus facilitating the process of extracting meaningful relational tuples, where each of them represents a separate fact from the input.

(6) **Most state-of-the-art Open IE approaches lack the expressiveness needed to properly represent complex assertions, resulting in incomplete, uninformative or incoherent extractions that have no meaningful interpretation or miss critical information asserted in the input sentence.** Even though there are no clear constraints on the relations to extract, previous work in the area of Open IE has mainly focused on the extraction of isolated relational tuples on a clausal or sentential level. State-of-the-art systems accomplish this task with high accuracy but ignore the cohesive nature of texts, where context

spread across clauses or sentences is needed to infer the true meaning of a statement. For instance, for a correct understanding of the example sentence from Table 1.1, it is important to distinguish between information asserted in the input and information that is only hypothetical or conditionally true. Here, the information contained in the statement *"caution with non water soluble contrast is mandatory"* depends on the pre-posed if-clause. However, many state-of-the-art Open IE approaches are not able to capture this relationship as they output only a loose arrangement of extracted tuples that ignore the context under which a proposition is complete and correct. Hence, in order to improve the interpretability of Open IE tuples, we suggest not only to extract isolated relations on clausal or sentential level, but to also incorporate the semantic context in the form of semantic relationships between extracted relational tuples.

More details on the problems faced by state-of-the-art syntactic TS and Open IE approaches will be discussed in our extensive literature review in Section 2.1 and Section 2.3. The related topics of text coherence and MR are investigated in Section 2.2 and Section 2.4.

1.4 Research Questions

Based on the problems in TS and Open IE that were identified in the previous section, we pose the following research questions:

Research Question 1. Is it possible to transform sentences that present a complex linguistic structure into a semantic hierarchy of minimal, self-contained propositions?

This research question can be detailed into the following research hypotheses, with three of them addressing the outcome of the splitting process (*Hypothesis 1.1. to Hypothesis 1.3.*) and two of them targeting the creation of the semantic hierarchy between the split components (*Hypothesis 1.4.* and *Hypothesis 1.5.*):

- [Hypothesis 1.1] **Syntactic Correctness**:

    A complex sentence C can be transformed into a sequence of simplified sentences $T_1, \ldots, T_n, n >= 2$, such that each simple sentence T_i is grammatically sound.

- [Hypothesis 1.2] **Minimality**:

  A complex sentence C can be transformed into a sequence of simplified sentences $T_1, \ldots, T_n$, $n >= 2$, such that each simple sentence T_i presents a minimal semantic unit, i.e. it cannot be further decomposed into meaningful propositions.

- [Hypothesis 1.3] **Semantic Correctness**:

  A complex sentence C can be transformed into a sequence of output sentences $T_1, \ldots, T_n$, $n >= 2$, such that the output sentences $T_1, \ldots, T_n$ convey all and only the information contained in the input.

- [Hypothesis 1.4] **Contextual Hierarchy**:

  For each pair of structurally simplified sentences T_i and T_j, it is possible to set up a contextual hierarchy between them, distinguishing core sentences that contain the key information of the input from contextual sentences that disclose less relevant, supplementary information.

- [Hypothesis 1.5] **Coherence Structure**:

  For each pair of structurally simplified sentences T_i and T_j, it is possible to preserve the coherence structure of complex sentences in the simplified output by identifying and classifying the semantic relationship that holds between two decomposed spans.

Research Question 2.   Does the generated output in the form of a semantic hierarchy of minimal, self-contained propositions support machine processing in terms of Open IE applications?

This research question can be subdivided into the following research hypotheses:

- [Hypothesis 2.1] **Coverage and Accuracy**:

  The minimal, self-contained propositions generated by our proposed TS approach improve the performance of state-of-the-art Open IE systems in terms of precision and recall.

- [Hypothesis 2.2] **Canonical Context-Preserving Representation of Relational Tuples**:

*The semantic hierarchy of minimal propositions generated by our discourse-aware TS
approach*

(a) *reduces the complexity of the relation extraction step, resulting in a simplistic
     canonical predicate-argument structure of the output, and*
(b) *allows for the enrichment of the extracted relational tuples with important
     contextual information that supports their interpretability.*

1.5   Contributions

This thesis makes a number of contributions to the fields of TS and Open IE by
critically analyzing existing approaches and proposing a new discourse-aware TS
approach with a subsequent relation extraction stage, generating semantically typed
relational tuples from complex source sentences.

The contributions of our work can be summarized as:

Contribution 1. We provide a critical overview of existing TS and Open IE
approaches and identify their main benefits and shortcomings, motivating our
research questions.

Contribution 2. We conduct a systematic analysis of syntactic phenomena involved
in complex sentence constructions whose findings are materialized in a small set
of 35 hand-crafted simplification rules.

Contribution 3. We propose a discourse-aware syntactic TS approach which transforms English sentences that present a complex linguistic structure into a semantic hierarchy of simplified sentences in the form of minimal, self-contained propositions. As a proof of concept, we develop a reference implementation (DISSIM).

Contribution 4. We perform a comprehensive evaluation against current syntactic
TS baselines, demonstrating that our reference implementation DISSIM outperforms the state of the art in structural TS both in an automatic evaluation and a
manual analysis across three TS datasets from two different domains.

Contribution 5. In an extrinsic evaluation of our discourse-aware TS approach,
we demonstrate its merits for downstream Open IE applications.

Contribution 6. We develop a supporting system for the relation extraction stage
(Graphene). It operates on the semantic hierarchy of simplified sentences with
the objective of extracting relational tuples within the semantic context in which
they occur.

Contribution 7. We introduce a supporting formal semantic representation of complex sentences based on the Resource Description Framework (RDF) standard.

It consists of a two-layered hierarchical representation of the input in the form of core sentences and accompanying contexts. Rhetorical relations are used to preserve the coherence structure of the input, resulting in a more informative output which is easier to interpret.

Contribution 8. We compile a large-scale sentence splitting corpus which is composed of 203K pairs of aligned complex source and simplified target sentences. This dataset may be useful for developing data-driven sentence splitting approaches that attempt to learn how to transform sentences with a complex linguistic structure into a fine-grained representation of short and simple sentences that present an atomic semantic unit.

The source code of both our reference TS implementation DISSIM[1] and our Open IE framework Graphene[2] is published online to support transparency and reproducibility of this work. MINWIKISPLIT, the sentence splitting corpus we compiled, is available online, too.[3]

1.6   Outline

The remainder of this thesis is organized as follows:

After introducing the target problem and the corresponding research questions that will be addressed in this work, **Chapter I** presents a critique of the state of the art in syntactic TS (Section 2.1) and Open IE (Section 2.3). We compare various previously proposed approaches and draw attention to their particular strengths and weaknesses, opening new research directions to be explored in this thesis (**Contribution 1**). In addition, we discuss approaches addressing the related problem of preserving coherence structures in texts (Section 2.2), as well as schemes for the semantic representation of sentences (Section 2.4).

Chapter II proposes a new discourse-aware TS approach (DISSIM) that transforms complex input sentences into a semantic hierarchy of minimal propositions, resulting in a novel hierarchical representation in the form of core sentences and accompanying contexts. Rhetorical relations are used to preserve the semantic context of the output (**Contribution 2** and **Contribution 3**). In addition, we present MINWIKISPLIT, a novel large-scale sentence splitting corpus of 203K pairs of complex source and aligned simplified target sentences in terms of a set of minimal

---

[1] https://github.com/Lambda-3/DiscourseSimplification

[2] https://github.com/Lambda-3/Graphene

[3] https://github.com/Lambda-3/minwikisplit

propositions, i.e. a sequence of sound, self-contained utterances with each of them presenting a minimal semantic unit that cannot be further decomposed into meaningful propositions (**Contribution 8**). This chapter addresses the first research question (**Research Question 1**).

Chapter III investigates how the proposed sentence splitting approach can be leveraged for the task of Open IE to facilitate the process of extracting a formal semantic representation from complex input sentences (**Contribution 7**). As a proof of concept, a supporting system is developed (Graphene) (**Contribution 6**). This chapter targets the second research question (**Research Question 2**).

Chapter IV reports on the results of the experiments that were conducted to evaluate the performance of our reference TS implementation DISSIM with regard to the two subtasks of (1) splitting and rephrasing syntactically complex input sentences into a set of *minimal propositions*, and (2) setting up a *semantic hierarchy* between the split components. With regard to the former, we compared the performance of our approach with state-of-the-art syntactic TS baselines on three commonly applied TS corpora from two different domains. To assess the latter, we performed a comparative analysis with the annotations contained in the RST-DT (Carlson et al., 2002) (**Contribution 4**). Finally, in an extrinsic evaluation of our discourse-aware TS approach, we demonstrate its use as a support for the Open IE process (**Contribution 5**). This chapter provides the experimental results backing the hypotheses proposed in **Research Question 1** and **Research Question 2**.

Chapter V revisits the main research questions and contributions of this thesis. It summarizes the experiments and most important findings of each chapter, comments on their potential impact on future TS and Open IE studies, and proposes new research avenues.

1.7   Associated Publications

Different aspects of this work were disseminated in previously published articles in proceedings of peer-reviewed international conferences:

1. **Christina Niklaus**, Matthias Cetto, André Freitas, and Siegfried Handschuh. 2019b. Transforming Complex Sentences into a Semantic Hierarchy. In *Proceedings of the 57th Annual Meeting of the Association for Computational Linguistics*, pages 3415–3427, Florence, Italy. Association for Computational Linguistics.
   This work proposes a method for transforming complex sentences into a semantic hierarchy of minimal propositions, thus introducing the key concepts of this

dissertation (see Chapter II). It describes an approach that uses a set of linguistically principled transformation patterns to recursively split and rephrase input sentences into shorter, syntactically simplified utterances. To avoid producing a set of disconnected components, both their semantic context and relations to other units are captured in terms of rhetorical relations, resulting in a novel hierarchical representation in the form of core sentences and accompanying contexts. The paper includes a comprehensive evaluation against state-of-the-art TS baselines, forming the basis of the experiments described in Section 15.1.

2. **Christina Niklaus**, Matthias Cetto, André Freitas, and Siegfried Handschuh. 2019a. DISSIM: A Discourse-Aware Syntactic Text Simplification Framework for English and German. In *Proceedings of the 12th International Conference on Natural Language Generation*, pages 504–507, Tokyo, Japan. Association for Computational Linguistics.

   This paper presents a reference implementation of our discourse-aware TS approach, DISSIM, which was developed as a proof of concept for transforming complex sentences into a semantic hierarchy of minimal propositions.

3. **Christina Niklaus**, André Freitas, and Siegfried Handschuh. 2019c. MINWIKISPLIT: A Sentence Splitting Corpus with Minimal Propositions. In *Proceedings of the 12th International Conference on Natural Language Generation*, pages 118–123, Tokyo, Japan. Association for Computational Linguistics.

   This paper introduces MINWIKISPLIT, a large-scale sentence splitting corpus of aligned pairs of complex source and simplified target sentences where each input sentence is broken down into a set of minimal propositions. This dataset is presented in Section 8.

4. **Christina Niklaus**, Matthias Cetto, André Freitas, and Siegfried Handschuh. 2018. A Survey on Open Information Extraction. In *Proceedings of the 27th International Conference on Computational Linguistics*, pages 3866–3878, Santa Fe, New Mexico, USA. Association for Computational Linguistics. (**Best Survey Paper**)

   This paper contributes an extensive survey on the task of Open IE, providing a critical summary of state-of-the-art approaches, resources and evaluation procedures. It lays the groundwork for the description of related work in the area of Open IE in Section 2.3. This publication received the award for best survey paper at the COLING Conference 2018.

5. Matthias Cetto, **Christina Niklaus**, André Freitas, and Siegfried Handschuh. 2018b. Graphene: Semantically-linked Propositions in Open Information Extraction. In *Proceedings of the 27th International Conference on Computational Linguistics*, pages 2300–2311, Santa Fe, New Mexico, USA. Association for Computational Linguistics.

This paper explores the relation extraction step of our TS—Open IE pipeline, described in Chapter III. It presents a method for leveraging the syntactically simplified sentences generated by our TS approach (see Chapter II) to improve the performance of state-of-the-art Open IE systems in terms of recall and precision. A thorough comparative evaluation against a set of Open IE baseline approaches was conducted, investigating the coverage and accuracy of the extracted relational tuples. This study forms the basis of our experiments from Section 15.3.

6. Matthias Cetto, **Christina Niklaus**, André Freitas, and Siegfried Handschuh. 2018a. Graphene: A Context-Preserving Open Information Extraction System. In *Proceedings of the 27th International Conference on Computational Linguistics: System Demonstrations*, pages 94–98, Santa Fe, New Mexico, USA. Association for Computational Linguistics.

This paper presents Graphene, a reference implementation of our proposed TS—Open IE pipeline. It was developed as a proof of concept for the extraction of semantically typed relational tuples from complex input sentences, with the objective of demonstrating that the semantic hierarchy of minimal propositions serves as an intermediate representation that facilitates the relation extraction process.

7. **Christina Niklaus**, Bernhard Bermeitinger, Siegfried Handschuh, and André Freitas. 2016. A Sentence Simplification System for Improving Relation Extraction. In *Proceedings of COLING 2016, the 26th International Conference on Computational Linguistics: System Demonstrations*, pages 170–174, Osaka, Japan. The COLING 2016 Organizing Committee.

This paper introduces the idea of using syntactic TS as a pre-processing step for improving the coverage and accuracy of the relational tuples extracted by state-of-the-art Open IE approaches. It presents a first prototype implementation of the discourse-aware TS approach developed in this thesis. Thus, the paper is the starting point of this dissertation.

Related Work

2

The following section presents a critical summary of previous work in the area of TS, focusing on methods that aim at simplifying the linguistic structure of sentences. We discuss the strengths and weaknesses of existing approaches, thereby demonstrating the effectiveness of our proposed context-preserving TS framework. In addition, in Section 2.2, we give a comprehensive overview of approaches that operate on the level of discourse by taking into account the coherence structure of texts. Section 2.3 then reviews approaches that were proposed to solve the task of Open IE. We highlight their limitations and provide a critique of commonly applied evaluation procedures. Finally, in Section 2.4, we briefly discuss the related topic of MR.

2.1 Text Simplification

TS aims to reduce the complexity of a text, while still retaining its meaning. It comprises both the modification of the vocabulary of the text (*lexical simplification*) and the modification of the structure of the sentences (*syntactic simplification*). Traditionally, TS has been reduced to the following four rewriting operations:

- *substituting* complex words or phrases with simpler synonyms,
- *reordering* sentence components,
- *deleting* less important parts of a sentence, and
- *splitting* a complex sentence into several simpler ones.

These transformations are executed with the objective of converting the input into a simplified output, while preserving its main idea and keeping it grammatically

© The Author(s), under exclusive license to Springer Fachmedien Wiesbaden GmbH, part of Springer Nature 2022
C. Niklaus, *From Complex Sentences to a Formal Semantic Representation using Syntactic Text Simplification and Open Information Extraction*,
https://doi.org/10.1007/978-3-658-38697-9_2

sound. Table 2.1 illustrates the four major types of TS operations mentioned above by means of example sentences.

Table 2.1 Simplification examples illustrating the major types of TS transformations. In the first two examples, uncommon words are replaced with a more familiar term (*bold*) and less relevant information is removed (*underlined*)[a]. In the third example, a long sentence is split into two simpler ones (*brackets*). At the same time, the original clause order is reversed, resulting in a cause-effect order that is easier to comprehend for readers with poor literacy (Siddharthan, 2014). In addition, a simple cue word is introduced (*italic*) to make the discourse relation between the split sentences explicit

Complex source	**Able-bodied** Muslims are **required** to visit Islam's holiest city, Mecca, at least once in their lifetime.
Simplified output	**Healthy** Muslims **need** to visit Mecca at least once in their lifetime.
Complex source	Alan Mathison Turing was a **British** mathematician, logician, cryptanalyst and computer scientist.
Simplified output	Alan Mathison Turing was an **English** mathematician and computer scientist.
Complex source (Siddharthan, 2006)	[The remaining 23,403 tons are still a lucrative target for growers] [*because* the U.S. price runs well above the world rate.]
Simplified output	[The U.S. price runs well above the world rate.] [*So* the remaining 23,403 tons are still a lucrative target for growers.]

[a]These examples are taken from the WikiLarge dataset (Zhang and Lapata, 2017). For details about this corpus, see Section 2.1.2.1.

2.1.1   Approaches to Simplification

In the following, we review research on TS, with the objective of understanding the benefits and shortcomings of the proposed approaches. We classify them based on the techniques that they use and compare their performance with respect to commonly applied TS metrics, as well as their transformation capabilities.

2.1.1.1 Rule-based Approaches

The task of structural TS was introduced by Chandrasekar et al. (1996), who manually define a set of rules to detect points where sentences may be split, such as relative pronouns or conjunctions. The rule patterns are based on chunking and dependency parse representations. Similarly, Siddharthan (2002) presents a pipelined architecture for a simplification framework that extracts a variety of clausal and phrasal components (appositive phrases, relative clauses, as well as coordinate and subordinate clauses) from a source sentence and transforms them into stand-alone sentences using a set of hand-written grammar rules based on shallow syntactic features. In Heilman and Smith (2010), the set of simplification rules, which are defined over phrasal parse trees, are augmented by rules for extracting participial modifiers, verb phrase modifiers and parentheticals, among others. This method is based on the idea of semantic entailment and presupposition.

More recently, Siddharthan and Mandya (2014) propose RegenT, a hybrid TS approach that combines an extensive set of 136 hand-written grammar rules defined over dependency tree structures for tackling seven types of linguistic constructs with a much larger set of automatically acquired rules for lexical simplification. Taking a similar approach, Ferrés et al. (2016) describe a linguistically-motivated rule-based TS approach called YATS, which relies on part-of-speech tags and syntactic dependency information to simplify a similar set of linguistic constructs, using a set of only 76 hand-crafted transformation patterns in total. A vector space model is applied to achieve lexical simplification. These two rule-based structural TS approaches primarily target reader populations with reading difficulties, such

Table 2.2 Comparison of the linguistic constructs that are addressed by the syntactic TS frameworks YATS, RegenT and our proposed approach DISSIM based on the number of hand-crafted grammar rules for simplifying these structures

LINGUISTIC CONSTRUCT	#RULES		
	YATS	**RegenT**	**DISSIM**
Correlatives	4	0	0
Coordinate clauses	10		1
Coordinate verb phrases		85 (lexicalised on conjunctions)	1
Adverbial clauses	12		6
Reported speech	0	14	4
Relative clauses	17	26	9
Appositions	1		2
Coordinate noun phrases	0	0	2
Participial phrases	0	0	4
Prepositional phrases	0	0	3
Adjectival and adverbial phrases	0	0	2
Lead noun phrases	0	0	1
Passive constructions	14	11	0
Total	76	136	35

as people suffering from dyslexia, aphasia or deafness. According to Siddharthan (2014), those groups most notably benefit from substituting difficult words, making discourse relations explicit, reordering sentence components and splitting long sentences that contain clausal constructs. Therefore, TS approaches that are aimed at human readers commonly represent hybrid systems that operate both on the syntactic and lexical level, with the decomposition of clausal elements representing the main focus of the structural simplification component.

In contrast to above-mentioned TS frameworks, our proposed TS approach does not address a human audience, but rather aims to **produce an intermediate representation that presents a simple and more regular structure that is easier to process for downstream Open IE applications.** For this purpose, we *cover a wider range of syntactic constructs*. In particular, our approach is not limited to breaking up clausal components, but also splits and rephrases a variety of phrasal elements, resulting in a much more fine-grained output where each proposition represents a minimal semantic unit that is composed of a normalized subject-verb-object structure (or simple variants thereof, see Table 4.1). Though tackling a larger set of linguistic constructs, our framework operates on a *much smaller set of only 35 manually defined rules*. A detailed comparison of the linguistic constructs that are handled by the two state-of-the-art rule-based syntactic TS systems YATS and RegenT and our reference TS implementation DISSIM is shown in Table 2.2.

Table 2.3 contrasts the output generated by RegenT and YATS on a sample sentence. As can be seen, RegenT and YATS break down the input into a sequence of sentences that present its message in a way that is easy to digest for human readers. However, the sentences are still rather long and present an irregular structure that mixes multiple semantically unrelated propositions, potentially causing problems for downstream Open IE tasks. On the contrary, our fairly aggressive simplification strategy that splits a source sentence into a large set of very short sentences is not well suited for a human audience and may in fact even hinder reading comprehension. However, we were able to demonstrate that the transformation process we propose improves the accuracy and coverage of the relational tuples extracted by state-of-the-art Open IE systems, serving as an example of a downstream NLP application that benefits from using our TS approach as a pre-processing step (see Section 15.3).

With LEXEV and EVLEX, Štajner and Glavaš (2017) present two TS frameworks that are able to perform sentence splitting, content reduction and lexical simplifications. These approaches are semantically-motivated, aiming to reduce the amount of descriptions in a text and keeping only event-related information. For this purpose, they make use of EVGRAPH (Glavaš and Štajner, 2015), a state-of-the-art event extraction system that relies on a supervised classifier to identify event mentions in a given sentence. In a second step, a set of 11 manually defined syntax-based extrac-

Table 2.3 Example comparing the output produced by the state-of-the-art rule-based syntactic TS frameworks RegenT and YATS with the simplification generated by our proposed approach DISSIM. Regarding DISSIM, the split sentences are represented in bold. Each of them is assigned a unique identifier *(#id)*, followed by a number that indicates the sentence's level of context. Moreover, the decomposed spans are connected to one another through rhetorical relations capturing their semantic relationship. More details on DISSIM's output format can be found in Section 12.2

SYSTEM	OUTPUT
Input	*The house was once part of a plantation and it was the home of Josiah Henson, a slave who escaped to Canada in 1830 and wrote the story of his life.*
RegenT	The house was once part of a plantation. And it was the home of Josiah Henson, a slave. This slave escaped to Canada in 1830 and wrote the story of his life.
YATS	The house was once part of a plantation. And it was the home of Josiah Henson. Josiah Henson was a slave who escaped to Canada in 1830 and wrote the story of his life.
DISSIM	#1　0　**The house was once part of a plantation.** 　　　LIST #2 #2　0　**It was the home of Josiah Henson.** 　　　ELABORATION #3 　　　LIST #1 #3　1　**Josiah Henson was a slave.** 　　　ELABORATION #4 　　　ELABORATION #6 #4　2　**This slave escaped to Canada.** 　　　TEMPORAL #5 　　　LIST　#6 #5　3　**This was in 1830.** #6　2　**This slave wrote the story of his life.** 　　　LIST #4

tion rules are applied to extract event-related arguments of four types: *agent, target, location* and *time*. Each identified event constitutes, together with its arguments, a new stand-alone simplified sentence. In this way, sentence splitting transformations are realized. Moreover, any information that does not belong to an event mention is discarded. Thus, content reduction is performed. Finally, an unsupervised approach based on word embeddings is used to carry out lexical simplification operations. In comparison to LEXEV and EVLEX, where *content reduction* plays a major role, we are interested in *preserving the full informational content* of an input sentence, as illustrated in Table 2.4.

Table 2.4 Example comparing the output produced by LexEv and EvLex with the simplification generated by our proposed approach DisSim

System	Output
Input	*"The amabassador's arrival has not been announced and he flew in complete secrecy," the official said.*
LexEv, EvLex	He arrived in complete secrecy.

DisSim	#1 0 **The ambassador's arrival has not been announced.**
	    LIST #2
	    ATTRIBUTION #3
	#2 0 **He flew in complete secrecy.**
	    LIST #1
	    ATTRIBUTION #3
	#3] 1 **This was what the official said.**

Lately, based on the work described in Evans (2011), Evans and Orăsan (2019) propose a sentence splitting approach that combines data-driven and rule-based methods. In a first step, they use a Conditional Random Field (CRF) tagger (Lafferty et al., 2001) to identify and classify signs of syntactic complexity, such as coordinators, *wh*-words or punctuation marks. Then, a small set of 28 manually defined rules for splitting coordinate clauses and a larger set of 125 hand-crafted grammar rules for decomposing nominally bound relative clauses is applied to rewrite long sentences into sequences of shorter single-clause sentences. Compared to Evans and Orăsan (2019)'s approach, which covers only two types of clausal constructs, our proposed discourse-aware TS framework targets a much wider range of sentence structures, while at the same time we rely on a set of rules that is considerably smaller.

Table 2.5 summarizes the performance of the rule-based TS approaches described above, based on a set of human ratings.[1] Note that the scores are not directly comparable, since the approaches were assessed on different corpora. In addition to that,

---

[1] In this setting, judges are asked to rate the quality of the simplified output produced by a TS system on a Likert scale that ranges from 1 to 5 with regard to three criteria: grammaticality, meaning preservation and simplicity. For further details, see Section 14.1.4.1.

Table 2.5 Performance of rule-based TS approaches as reported by their authors. A (*) indicates a test set that was compiled solely for the purpose of evaluating the proposed system. Details on the remaining standard corpora can be found in Section 2.1.2. A (†) denotes the 95% confidence interval

Approach	#rules	Test Corpus	#raters	Grammaticality	Meaning Preservation	Simplicity
Heilman and Smith (2010)	28	encyclopedia articles*	2	4.75	4.67	–
RegenT	136	PWKP	5	4.04	4.60	3.74
YATS	76	PWKP	8	4.58	3.76	3.93
LexEv	11	News*	2	4.23	3.57	4.49
	11	Wiki*	2	4.35	3.58	4.30
EvLex	11	News*	2	4.25	3.55	4.49
	11	Wiki*	2	4.35	3.54	4.25
Evans and Orăsan (2019)	153	Health, News & Literature*	5	[3.79, 4.05]†	[3.72, 4.02]†	[4.03, 4.25]†

the evaluation was carried out by different annotators on the basis of slightly varying annotation guidelines.

2.1.1.2 Approaches based on Monolingual Statistical Machine Translation

Other approaches treat TS as a monolingual MT task, with the original sentences as the source and the simplified sentences as the target language. In statistical machine translation (SMT), the goal of a model is to produce a translation e in the target language when given a sentence f in the source language. Various different approaches for implementing the translation model have been proposed so far. One of them is phrase-based machine translation (PBMT), which uses phrases, i.e. sequences of words, as the basic unit of translation. Specia (2010) was the first to apply this approach for the task of TS. Her model is based on the PBMT toolkit Moses (Koehn et al., 2007), which is used without any adaptations for carrying out TS transformations. It mostly executes lexical simplifications and simple structural rewritings, while more complex syntactic transformations such as sentence splittings cannot be performed. Coster and Kauchak (2011a) extend this approach to allow complex phrases to be aligned to NULL, thus implementing deletion operations ("Moses-Del"). Wubben et al. (2012) also use Moses, but add a post-processing stage where the output sentences are re-ranked according to their dissimilarity from the input, with the goal of encouraging transformation operations ("PBMT-R"). The resulting simplifications are limited to paraphrasing rewrites. In short, PBMT models are able to perform substitutions, short distance reorderings and deletions. However, they fail to learn more sophisticated operations like sentence splitting that may require more information on the structure of the sentences and relationships between their components (Alva-Manchego et al., 2020).

Another SMT-based approach is syntax-based machine translation (SBMT). Here, the fundamental unit for translation is no longer a phrase, but syntactic components in parse trees. This allows to extract more informed features, based on the structure of parallel trees of source and target sentences (Alva-Manchego et al., 2020). An example of an SBMT approach is "TSM" (Zhu et al., 2010), which makes use of syntactic information on the source side in order to perform four rewriting operations, namely reordering, splitting, deletion and substitution. For estimating the translation model, the method traverses the parse tree of the original sentence from top to bottom, extracting features for each possible transformation. It is trained on the PWKP corpus (see Section 2.1.2) using the Expectation Maximization algorithm proposed in Yamada and Knight (2001). In the evaluation, it demonstrates good performance for word substitution and sentence splitting. Focusing on lexical simplification, Xu et al. (2016) trained an SBMT model on a large-scale paraphrase dataset (PPDB) (Ganitkevitch et al., 2013). It optimizes an SBMT framework (Post et al., 2013) with rule-based features and tuning metrics specifically tailored for the task of lexical simplification (BLEU, FKBLEU and SARI). Hence, it is limited to lexical simplification operations, whereas "TSM" is capable of performing more complex syntactic transformations, such as splits.

Table 2.6 summarizes the performance of the SMT-based TS approaches described above, using a set of commonly applied TS metrics.[2] Note that the values are not directly comparable, since different corpora were used for testing the proposed models.

---

[2] Further details on the metrics of BLEU and SARI can be found in Section 14.1.3.2. Flesch-Kincaid Grade Level (FKGL) (Kincaid et al., 1975) is a widely applied readability metric, measuring how easy a text is to understand for human readers. It is based on average sentence length and average number of syllables per word. The result is a number that corresponds with a grade level in the United States.

Table 2.6 Performance of SMT-based TS approaches as reported by their authors (Alva-Manchego et al., 2020)

Approach	Train Corpus	Test Corpus	BLEU ↑	FKGL ↓	SARI ↑
Specia (2010)	Own[a]	Own[a]	59.87	–	–
Moses-Del	C&K-1	C&K-1	60.46	–	–
PBMT-R	PWKP	PWKP	43.00	13.38	–
TSM	PWKP	PWKP	38.00	–	–
SBMT (PPDB+BLEU)	PWKP Turk	PWKP Turk	99.05	12.88	26.05
SBMT (PPDB+FKBLEU)	PWKP Turk	PWKP Turk	74.48	10.75	34.18
SBMT (PPDB+SARI)	PWKP Turk	PWKP Turk	72.36	10.90	37.91

[a] Corpus of 104 manually simplified Portuguese newswire articles, amounting to 4,483 original sentences and their corresponding simplifications.

2.1.1.3 Approaches based on the Induction of Synchronous Grammars

Moreover, TS approaches that are based on the induction of synchronous grammars were proposed. Here, TS is modelled as a tree-to-tree rewriting problem. Typically, a two step process is carried out. First, parallel corpora of aligned pairs of complex source and simplified target sentences are used to extract a set of tree transformation rules. Then, the approaches learn how to select which rules to apply to unseen sentences in order to generate the best simplified output (*decoding*) (Alva-Manchego et al., 2020).

The approach described in Woodsend and Lapata (2011) ("QG+ILP") uses Quasisynchronous Grammars (QGs) (Smith and Eisner, 2006) to induce rewrite rules that target sentence splitting, word substitution and deletion operations. In a first step, simplification rules are extracted from constituent alignments between pairs of trees of complex source and simplified target sentences. Next, Integer Linear Programming (ILP) is used to find the best candidate simplification. During this process, the parse tree of the source sentence is traversed from top to bottom, applying at each node all the transformation rules that match. The objective function of the ILP involves a penalty that favors more common transformation rewrites and tries to reduce the number of words and syllables in the resulting simplified output.

Another grammar-induced TS approach is proposed in Paetzold and Specia (2013) ("T3+Rank"). To extract a set of simplification rules, it makes use of T3 (Cohn and Lapata, 2009), an abstractive sentence compression model that employs

Synchronous Tree Substitution Grammars (STSGs) (Eisner, 2003) to perform deletion, reordering and substitution operations. It starts by mapping the word alignment
of complex-simple sentence pairs into a constituent level alignment between their
source and target trees by adapting the alignment template method of Och and Ney
(2004). Next, the alignments are generalized, i.e. the aligned nodes are replaced with
links, using a recursive algorithm whose goal is to find the minimal most general set
of transformation rules. During decoding, candidate simplifications are then ranked
according to their perplexity in order to determine the best simplifications.

Furthermore, Feblowitz and Kauchak (2013) ("SimpleTT") present an approach
similar to the one described in Cohn and Lapata (2009), modifying the rule extraction process to reduce the number of candidate simplification rules that need to be
generalized. During decoding, the model tries to match the more specific rules up
to the most general, thus generating the 10,000 most probable simplifications, from
which the best one is determined using a log-linear combination of features, such
as rule probabilities and output length.

Table 2.7 Performance of grammar-induced TS approaches as reported by their authors. A
(*) indicates a training/test set different from the standard (Alva-Manchego et al., 2020)

Approach	Train Corpus	Test Corpus	BLEU ↑	FKGL ↓
QG+ILP	AlignedWL	PWKP	34.00	12.36
	RevisionWL	PWKP	42.0	10.92
T3+Rank	C&K-1*	C&K-1*	34.2	–
SimpleTT	C&K-1	C&K-1	56.4	–

Alva-Manchego et al. (2020) claim that due to their pipeline architecture,
grammar-induced-based approaches offer more flexibility on how the rules are
learned and how they are applied, as compared to end-to-end approaches. Even
though Woodsend and Lapata (2011) were the only ones to explicitly model sentence splitting, the other approaches could be modified in a similar way to incorporate splitting operations. Table 2.7 presents an overview of the performance of
the grammar-induced TS approaches mentioned above. Since different test corpora
were applied, the scores are not directly comparable.

2.1.1.4 Semantics-Assisted Approaches

Arguing that sentence splitting is semantics-driven, Narayan and Gardent (2014)
propose HYBRID, a TS approach that takes semantically-shared elements as the
basis for splitting and rephrasing a sentence. It is based on the assumption that splits

occur when an entity takes part in two or more distinct events described in a single sentence. For instance, in the example depicted in Table 2.8, the split is on the entity "Donald Trump", which is involved in two events, namely "being inaugurated as president" and "moving his businesses into a revocable trust".

Table 2.8 Semantically-shared elements as the basis for splitting a sentence

Complex source	Before being inaugurated as president, *Donald Trump* moved his businesses into a revocable trust.
Simplified output	*Donald Trump* was inaugurated as president. *Donald Trump* moved his businesses into a revocable trust.

HYBRID combines semantics-driven deletion and sentence splitting with PBMT-based substitution and reordering. To determine potential splitting points in a given sentence, this method first uses Boxer (Curran et al., 2007) to generate a semantic representation of the input in the form of Discourse Representation Structure (DRS) (Kamp, 1981) (see Section 2.4). Splitting candidates are pairs of events that share a common thematic role, e.g. *agent* and *patient* (see Section 2.4). The probability of a candidate being split is determined by the thematic roles associated with it. The split with the highest probability is chosen, and the decomposed spans are rephrased by completing them with the shared elements in order to reconstruct grammatically sound sentences. Deletion, too, is directed by thematic roles information, based on the assumption that constituents that are related to a predicate by a core thematic role (e.g., *agent* and *patient*) represent mandatory arguments and, thus, should not be deleted. Finally, the probabilities for substitution and reordering are determined by a PBMT system (Koehn et al., 2007).

In Narayan and Gardent (2016) ("UNSUP"), the authors present another semantics-assisted TS approach that does not require aligned original-simplified sentence pairs to train a TS model. Analogous to the method presented in Narayan and Gardent (2014), they apply Boxer to produce a semantic representation of the input, which is used to identify the events and predicates contained in it. To determine whether and where to split, the maximum likelihood of the sequences of thematic role sets resulting from each possible split are estimated. The underlying probabilistic model only relies on data from Simple English Wikipedia (SEW) (see Section 2.1.2). Phrasal deletion is carried out using ILP. Finally, the pipeline incorporates a context-aware lexical simplification component (Biran et al., 2011) that learns paraphrasing rules from aligned articles of English Wikipedia (EW) and SEW.

Table 2.9 Examples comparing the output produced by semantics-assisted TS models with the simplifications generated by the reference implementation DISSIM of our proposed discourse-aware TS approach

SYSTEM	OUTPUT
Input	*The tarantula, the trickster character, spun a black cord and, attaching it to the ball, crawled away fast to the east, pulling on the cord with all his strength.*
DSS	the tarantula the trickster character spun a black cord. attaching it to the ball. character crawled away fast to the east. character pulling on the cord with all his strength.
SENTS	the character has a character on the rope, and the player can see the ball.
HYBRID	the tarantula, the trickster character, a black spun cord, and it attaching, crawled, pulling all.
DISSIM	#1 0 **The tarantula spun a black cord.** LIST #2 LIST #3 ELABORATION #4 #2 0 **The tarantula crawled away fast to the east.** LIST #1 LIST #3 ELABORATION #4 #3 0 **The tarantula was pulling on the cord with all his strength.** LIST #1 LIST #2 ELABORATION #4 #4 1 **The tarantula was the trickster character.**

(a) Example 1.

(Continued)

Table 2.9 (Continued)

SYSTEM	OUTPUT
Input	*Their granddaughter Hélène Langevin-Joliot is a professor of nuclear physics at the University of Paris, and their grandson Pierre Joliot, who was named after Pierre Curie, is a noted biochemist.*
DSS	their granddaughter Hélène Langevin Joliot is a professor of nuclear physics at the university of Paris. Joliot was named after Pierre curie.
SENTS	their granddaughter Hélène Langevin Joliot is a professor of nuclear physics.
HYBRID	their granddaughter hélène langevin-joliot is a professor of physics, and their grandson pierre joliot, who was named, is a biochemist.

DISSIM

#1 0 **Hélène Langevin-Joliot is a professor of nuclear physics at the University of Paris.**

ELABORATION #2

LIST #3

#2 1 **Hélène Langevin-Joliot is their granddaughter.**

#3 0 **Pierre Joliot is a noted biochemist.**

LIST #1

ELABORATION #3

ELABORATION #5

#4 1 **Pierre Joliot is their grandson.**

#5 1 **Pierre Joliot was named after Pierre Curie.**

ELABORATION #6

#6 2 **Pierre Joliot is their grandson.**

(b) Example 2.

With Direct Semantic Splitting (DSS), Sulem et al. (2018c) describe a further semantics-based structural TS framework that follows a similar approach, making use of the TUPA parser presented in Hershcovich et al. (2017). It generates a semantic representation of the input sentence in the form of Universal Conceptual Cognitive Annotation (UCCA) (Abend and Rappoport, 2013) (see Section 2.4), which supports the direct decomposition of a sentence into individual semantic elements. The authors specify two semantic rules for sentence splitting that are conditioned on UCCA's event categories. The first rule targets parallel events ("Parallel Scenes"), which are extracted and separated into different sentences. The second rule aims at events that provide additional information about an established entity ("Elaborator Scenes"), which, too, are extracted from the source sentence together with the phrase to which they refer. In addition to DSS, Sulem et al. (2018c) propose SENTS, a hybrid system that adds a lexical simplification component to DSS by running the split sentences through the NMT-based system proposed in Nisioi et al. (2017) (see Section 2.1.1.5).

In contrast to SMT-based and grammar-induced TS approaches, where sentence splitting transformations play only a subordinate role, the semantics-assisted TS models described above are largely motivated by this type of rewriting operation. With the objective of decomposing complex source sentences into minimal semantic units, semantics-aided TS methods follow a very similar goal as compared to our syntactic TS approach. However, they generally suffer from a *poor grammaticality and meaning preservation* of the simplified output, as shown by the example simplifications in Table 2.9.

Table 2.10 provides a summary of the performance of the semantics-assisted TS approaches presented above.

Table 2.10 Performance of semantics-assisted TS approaches as reported by their authors (Alva-Manchego et al., 2020)

Approach	Train Corpus	Test Corpus	BLEU ↑	SARI ↑
Hybrid	PWKP	PWKP	53.60	–
UNSUP	–	PWKP	38.47	–
DSS	–	PWKP Turk	76.57	36.76
SENTS	EW-SEW	PWKP Turk	57.71	31.90

2.1.1.5 Neural Sequence-to-Sequence Approaches

Recent improvements in TS were achieved by the use of neural machine translation (NMT)-inspired approaches, where TS is modelled as a sequence-to-sequence (Seq2Seq) problem that is commonly addressed with an attention-based encoderde-coder architecture (Bahdanau et al., 2015; Cho et al., 2014; Sutskever et al., 2014). Most models rely on a Recurrent Neural Network (RNN) architecture with Long Short-Term Memory (LSTM) units (Hochreiter and Schmidhuber, 1997).

With Neural Text Simplification (NTS), Nisioi et al. (2017) introduced the first neural Seq2Seq TS approach, using an encoder-decoder architecture with an attention mechanism provided by OpenNMT (Klein et al., 2017). The resulting model is able to simultaneously perform lexical simplifications and content reduction. Alva-Manchego et al. (2017), too, explore this standard Seq2Seq architecture, but employ the implementation of Nematus (Sennrich et al., 2017). In Scarton and Specia (2018), the authors propose to augment the neural architecture by enriching the encoder's input with information about the target audience and the predicted TS transformations that need to be performed ("target-TS"). Replacing LSTMs with Neural Semantic Encoders (NSEs) (Munkhdalai and Yu, 2017), Vu et al. (2018) use a slightly different network architecture which allows to capture more context information while encoding the current token, since an NNSE has access to all the tokens in the input sequence, whereas an LSTM only relies on the previous hidden state ("NSELSTM").

Adhering to the standard RNN-based architecture, further research explores *alternative learning algorithms* for training TS models. For instance, Zhang and Lapata (2017) present DRESS, which applies the standard attention-based Seq2Seq architecture as an agent within a Reinforcement Learning (RL) architecture. In this setting, a TS model is trained in an end-to-end manner using TS-specific metrics as reward. Moreover, Guo et al. (2018) suggest a Multi-Task Learning (MTL) framework with TS as the main task, while paraphrase generation and entailment generation serve as auxiliary tasks to improve the TS model's performance ("POINTERCOPY+MTL"). In addition, they augment the standard encoder-decoder architecture with a pointer-copy mechanism (See et al., 2017; Gu et al., 2016), which allows to decide at decoding time whether to copy a token from the input or to generate one.

In contrast to previously proposed NMT-based TS approaches that attempt to learn simplification rewrites solely from the training datasets, Zhao et al. (2018) propose to *incorporate external knowledge* in the form of human-curated paraphrasing rules from Simple PPDB (Pavlick and Callison-Burch, 2016) ("DMASS+DCSS"). The authors argue that due to the relatively small size of TS datasets, the models have difficulty in capturing less frequent transformations and thus do not generalize well. Hence, claiming that external knowledge may help in learning rules that are not frequently observed in the underlying training data, paraphrasing rules are included in their approach. In addition, they move away from RNN-based architectures and instead use a Transformer architecture (Vaswani et al., 2017).

In Alva-Manchego et al. (2017), the TS task is modelled as a *Sequence Labeling (SL) problem* where TS rewrites are identified at the level of individual words or phrases. MASSAlign (Paetzold et al., 2017) is applied to automatically generate token-level annotations from which an LSTM learns to predict deletion and substitution operations. During decoding, words that are labelled to be deleted are not included in the output, while the system proposed in Paetzold and Specia (2017) is used for lexical simplification.

More recently, with UNTS, Surya et al. (2019) present an *unsupervised end-to-end TS approach*. It is motivated by the goal of designing an architecture that allows

Table 2.11 Performance of Seq2Seq TS approaches as reported by their authors (Alva-Manchego et al., 2020)

Approach	Train Corpus	Test Corpus	BLEU ↑	FKGL ↓	SARI ↑
NTS	EW-SEW	PWKP Turk	84.51	–	30.65
target-TS	Newsela	Newsela	64.78	5.47	45.41
DRESS	WikiSmall	PWKP	34.53	7.48	27.48
	WikiLarge	PWKP Turk	77.18	6.58	37.08
	Newsela	Newsela	23.21	4.13	27.37
NSELSTM- B	WikiSmall	PWKP	53.42	–	17.47
(optimized for	WikiLarge	PWKP Turk	92.02	–	33.43
BLEU)	Newsela	Newsela	26.31	–	27.42
NSELSTM- S	WikiSmall	PWKP	29.72	–	29.75
(optimized for	WikiLarge	PWKP Turk	80.43	–	36.88
SARI)	Newsela	Newsela	22.62	–	29.58
POINTERCOPY	WikiSmall	PWKP	29.70	6.93	28.24
+MTL	WikiLarge	PWKP Turk	81.49	7.41	37.45
	Newsela	Newsela	11.86	1.38	32.98
DMASS+DCSS	WikiLarge	PWKP Turk	–	8.04	40.45
	WikiLarge	Newsela	–	5.17	27.28
SL	Newsela	Newsela	41.37	–	31.29
UNTS	Own (Wikipedia)	PWKP Turk	74.24	–	33.8

to train TS models for languages and domains that lack large-scale resources of aligned pairs of complex source and simplified target sentences. It is based on an autoencoder architecture that uses a shared encoder and two dedicated decoders: one for generating complex sentences and one for producing simple sentences. The model also relies on a discriminator module that determines if a given context vector sequence is close to one extracted from simple sentences in the dataset, as well as a classifier module whose task is to bring about diversification in the generated output.

Up to that point, research in TS has been mostly focused on developing approaches that generate a single generic simplification for a given source text, with no possibility to adapt the output to the specific needs of various target groups. However, different audiences require different types of simplifications (Siddharthan, 2014). Hence, aiming to close this gap, the first attempts at *controllable sentence simplification* were presented. For instance, with ACCESS (**A**udience-**C**entric **S**entence

Table 2.12 Example comparing the output produced by a Seq2Seq sentence splitting model with the simplifications generated by the reference implementation DISSIM of our proposed discourse-aware TS approach

SYSTEM	OUTPUT
Input	*Diller suddenly sold the stations to Univision, which converted them into Spanish language outlets, and while some affiliated with Univision, most joined Telefutura, its new sister network that launched in January 2002.*
Copy512	Diller suddenly sold the stations to Univision, which converted them into Spanish language outlets. While some affiliated with Univision, most joined Telefutura, its new sister network that launched in January 2002.
DISSIM	#1 0 **Diller suddenly sold the stations to Univision.** ELABORATION #2 LIST #4 #2 1 **Univision converted them into Spanish language outlets.** #3 0 **Some affiliated with Univision.** ELABORATION #2 CONTRAST #4 #4 0 **Most joined Telefutura.** LIST] #1 CONTRAST #3 ELABORATION #5 #5 1 **Telefutura was its new sister network.** ELABORATION #6 #6 2 **This new sister network launched in 2002.**

Simplification), Martin et al. (2020) propose a TS approach which modifies a generic
Seq2Seq model to control various qualities of the output text. For this purpose, the
authors first identify a number of attributes that capture important features of the
simplification process, such as the amount of compression, paraphrasing, lexical and
syntactic complexity, and then find explicit control tokens to represent each of them.
Next, using these tokens, a Seq2Seq model is trained with the objective of learning to
associate different aspects of the simplification process with the control tokens. As a
result, by carefully choosing appropriate values for the attributes, the model can be
adapted more precisely to the type of simplification needed by the respective user.
Moreover, Dong et al. (2019) present EditNTS, a neural programmer-interpreter
approach (Reed and de Freitas, 2016) for the task of TS, where simplifying edit
operations (*add, delete, keep*) are predicted on the words of a complex input sentence by the programmer and then executed into simplified tokens by the interpreter.
By adjusting the loss weights on the edit operations, the model can prioritize simplification operations for different tasks, thus offering control over the output length,
the number of words that are copied from the input, as well as the number of new
words that are added to the simplified output sentence. Kriz et al. (2019), too, support personalized TS by extending a generic Seq2Seq framework. For this purpose,
they incorporate word complexities into the model's loss function during training
in order to encourage it to choose simpler words, and generate a large set of diverse
candidate simplifications at test time, which are reranked based on a scoring function
to select the simplification that promotes fluency, adequacy and simplicity. Consequently, by changing the loss and scoring function, the model offers control over
the simplified output.

An important limitation of the NMT-based TS systems presented above is that
they *do not address sentence splitting transformations*, potentially due to the scarcity
of this type of rewriting operation in the training corpora (Sulem et al., 2018c).
Table 2.11 summarizes the performance of the above-mentioned neural Seq2Seq
approaches, using a set of TS metrics.

The work of Narayan et al. (2017) builds on this observation. The authors argue
that established TS corpora, such as PWKP (Zhu et al., 2010), C&K-1 (Coster
and Kauchak, 2011b), Newsela (Xu et al., 2015) and WikiLarge (Zhang and Lapata,
2017) (see Section 2.1.2), are ill suited for learning how to decompose sentences into
shorter, syntactically simplified components, as they contain only a small number
of split examples. Therefore, they propose a new corpus, WebSplit. It is the first TS
dataset that explicitly addresses the task of sentence splitting, while abstracting away
from other TS transformations, such as paraphrasing or deletion operations. This
dataset is composed of over one million tuples that map a single complex sentence
to a sequence of structurally simplified sentences expressing the same meaning

(for details, see Section 2.1.2). Introducing a new task called "Split-and-Rephrase" that focuses on splitting a sentence into several shorter ones, Narayan et al. (2017) describe a set of Seq2Seq models trained on the WebSplit corpus for breaking up complex input sentences into shorter components that present a simpler structure.

Aharoni and Goldberg (2018) further explore this idea, augmenting the presented neural models with a copy mechanism (See et al., 2017; Gu et al., 2016). Though outperforming the models of Narayan et al. (2017), they still perform poorly compared to previous state-of-the-art rule-based syntactic TS approaches. In addition, Botha et al. (2018) compile another sentence splitting corpus based on Wikipedia edit histories (see Section 2.1.2). When training the best-performing model of Aharoni and Goldberg (2018) on this corpus, they achieve a strong improvement over prior best results. However, as pointed out in Table 2.12 ("Copy512"), Botha et al. (2018)s' approach exhibits a *strong conservatism* since each input sentence is split in exactly two output sentences only, due to the uniform use of a single split per source sentence in the training set. Consequently, the resulting simplified sentences are still comparatively long and complex, mixing multiple, potentially semantically unrelated propositions that are difficult to handle for downstream Open IE tasks. Though, as opposed to the semantics-assisted approaches presented above, the neural Seq2Seq sentence splitting approaches commonly succeed in producing split sentences that are grammatically sound.

Table 2.13 Performance of Seq2Seq sentence splitting approaches as reported by their authors. A (*) indicates the data split proposed in Aharoni and Goldberg (2018), while a (Δ) indicates a scaled-up re-release of the original WebSplit corpus by Narayan et al. (2017) that, too, addresses the issues identified in Aharoni and Goldberg (2018). A ($^{\#}$) signifies a score that was not reported in the paper, but calculated by us. #S/C refers to the average number of simplified sentences per complex input sentence. #T/S represents the average number of tokens per simplified output sentence

Approach	Train Corpus	Test Corpus	BLEU ↑	#S/C	#T/S
Narayan et al. (2017)	WebSplit*	WebSplit*	6.68	2.44	10.23
Aharoni and Goldberg (2018)	WebSplit*	WebSplit*	24.97	2.87	10.04
Botha et al. (2018)	WebSplit v1.0$^{\Delta}$	WebSplit v1.0$^{\Delta}$	60.4	2.0	11.0
	WikiSplit	WikiSplit	76.0	2.08$^{\#}$	16.55$^{\#}$

Table 2.13 provides an overview of the performance of the encoder-decoder models that were trained to explicitly address sentence splitting rewrites.

2.1.1.6 Summary

In the previous sections, we collated the research that has been conducted so far in the area of TS, discussing the benefits and shortcomings of each approach. Following the work of Alva-Manchego et al. (2020), Table 2.14 lists the surveyed TS approaches and groups them according to the techniques that they explore for simplifying a given input sentence. In addition, we report the simplification transformations that each of them can perform, considering the four traditional TS rewriting operations (*substitution, splitting, deletion* and *reordering*), as acknowledged by their authors.

Table 2.14 Summary of the surveyed TS approaches, grouped according to the techniques that they explore. The right column specifies the simplification transformations that they can perform, considering the four traditional TS rewriting operations (*substitution, splitting, deletion* and *reordering*), as acknowledged by their authors (Alva-Manchego et al., 2020)

Model	Approach	Transformations
Heilman and Smith (2010)	28 rules based on phrasal parse tree structures	**Split**
Siddharthan and Mandya (2014)	136 rules defined over dependency tree structures (syntactic) and a grammar-induced approach (lexical)	SUB, **Split**, DEL, REORD
Ferrés et al. (2016)	76 rules based on part-of-speech tags and dependency tree structures (syntactic) and a vector space model (lexical)	SUB, **Split**, REORD
Štajner and Glavaš (2017)	Event detection for splitting and content reduction (syntactic) and an unsupervised approach (lexical)	SUB, **Split**, DEL
Evans and Orăsan (2019)	CRF tagger to identify signs of complexity and 153 rules defined over part-of-speech tags and lexical information	**Split**

(a) Rule-based TS approaches.

(Continued)

Table 2.14 (Continued)

Model	Approach	Transformations
Specia (2010)	PBMT (Moses)	SUB, REORD
Coster and Kauchak (2011a)	PBMT (Moses) + Deletion	SUB, DEL, REORD
Wubben et al. (2012)	PBMT (Moses) + Dissimilarity Reranking	SUB
Zhu et al. (2010)	SBMT	SUB, **Split**, DEL, REORD
Xu et al. (2016)	SBMT (Joshua) + PPDB + SARI/BLEU/FKBLEU Optimization	SUB

(b) SMT-based TS approaches.

Model	Approach	Transformations
Woodsend and Lapata (2011)	QGs + ILP	SUB, **Split**, DEL, REORD
Paetzold and Specia (2013)	STSGs + Perplexity Ranking	SUB, DEL, REORD
Feblowitz and Kauchak (2013)	STSGs + Log-linear Ranking	SUB, DEL, REORD

(c) Grammar-induced-based TS approaches.

Model	Approach	Transformations
Narayan and Gardent (2014)	DRS-based semantic representation (Boxer) + PBMT	SUB, **Split**, DEL, REORD
Narayan and Gardent (2016)	DRS-based semantic representation (Boxer) + ILP + context-aware lexical simplification	SUB, **Split**, DEL
Sulem et al. (2018c)	UCCA-based semantic representation (TUPA) + Seq2Seq (RNN)	SUB, **Split**

(d) Semantics-assisted TS approaches.

(Continued)

Table 2.14 (Continued)

Model	Approach	Transformations
Nisioi et al. (2017)	Seq2Seq (RNN)	Sub, Del, Reord
Alva-Manchego et al. (2017)	SL	Sub, Del
Zhang and Lapata (2017)	Seq2Seq (RNN) + RL	Sub, Del
Scarton and Specia (2018)	Seq2Seq (RNN) + Naive Bayes classifier	Sub, **Split**
Vu et al. (2018)	Seq2Seq (RNN) with NSE	Sub, Del
Guo et al. (2018)	Seq2Seq (RNN) + MTL	Sub, Del, Reord
Zhao et al. (2018)	Seq2Seq (Transformer) + SPPDB	Sub, Del
Surya et al. (2019)	Seq2Seq (autoencoder) + Discriminator + Classifier	Sub, **Split**, Del
Dong et al. (2019)	neural programmer-interpreter	Sub, Del
Kriz et al. (2019)	Seq2Seq (LSTM) + complexity-weighted loss function + candidate reranking based on fluency, adequacy and simplicity	Sub, Del
Martin et al. (2020)	Seq2Seq (Transformer) + explicit control tokens	Sub, Del, **Split**, Reord
Narayan et al. (2017)	Seq2Seq (RNN)	Sub, **Split**, Reord
Aharoni and Goldberg (2018)	Seq2Seq (RNN)	Sub, **Split**, Reord
Botha et al. (2018)	Seq2Seq (RNN)	**Split**

(e) Neural Seq2Seq TS approaches.

As Table 2.14b shows, *SMT-based methods* are capable of performing paraphrasing operations as well as simple structural transformations in terms of deleting and reordering sentence components. However, they commonly fail to carry out more complex syntactic rewritings, such as sentence splitting, unless they are explicitly modelled using structural information.

In contrast, *grammar-based approaches*, which extract a set of tree transformation rules, can model sentence splits more naturally. The process of selecting the set of rules to execute as well as their application order to generate the best simplification output is complex, though (Alva-Manchego et al., 2020). In fact, as can be seen in Table 2.14c, most TS approaches that are based on the induction of synchronous grammars ignore splitting transformations when simplifying a given input sentence.

As opposed to SMT-based and grammar-induced TS approaches, where sentence splitting transformations are largely neglected, the *semantics-assisted* TS models are to a great extent motivated by this type of rewriting operation (see Table 2.14d). Based on a semantic representation of the input, they aim to decompose complex sentences into individual semantic units, therefore putting a special focus on splitting operations. However, our examinations revealed that semantics-assisted TS approaches are prone to produce a simplified output that is not grammatically sound and does not preserve the meaning of the input (see Section 15.1.2.1).

The use of *neural approaches* gave rise to recent improvements in TS (see Table 2.14e). Most systems are variants of Seq2Seq models adapted from the field of MT, operating in an end-to-end fashion by extracting features automatically. They learn to simplify sentences implicitly, as a byproduct of the fact that they are trained on complex-simple sentence pairs (Dong et al., 2019). Consequently, one of the main problems with applying generic Seq2Seq models for the task of TS is that they tend to copy directly from the input, since this is the most common operation in TS, resulting in outputs that are relatively long and complex, often even identical to the input (Kriz et al., 2019).

Unlike Seq2Seq approaches, which can be considered as black boxes with respect to which simplification transformations should be applied to a given sentence (Alva-Manchego et al., 2020), *rule-based TS approaches* (see Table 2.14a) allow for a more modular approach that first determines which simplifications should be performed and then decides how to handle each transformation independently. In fact, TS systems that are based on a manually defined set of transformation rules offer more flexibility regarding the specification of the simplification rewrites, allowing to focus on a selected set of linguistic constructs to simplify. In that way, sentence splitting transformations cannot only be explicitly addressed, but also the granularity of the splits can be determined as required. Accordingly, rule based approaches support a wide range of syntactic transformation operations and provide a high level of flexibility to adapt the output to the particular needs of the respective target group or system. Though costly to produce, state-of-the-art structural TS approaches are to a large extent rule-based (Štajner and Glavaš, 2017).

2.1.2   Data Resources for Simplification

Many of the TS approaches presented above are data-driven, i.e. they learn to simplify from examples in corpora. These instances consist of aligned pairs of original
sentences and their corresponding simplified versions. A variety of parallel corpora
were compiled so far, containing different types of sentence alignments:

- **1-to-1**: An original sentence is aligned to one simplified sentence. This type of
  alignment allows for learning how to perform *substitution, reordering, deletion,
  insertion* and *identity transformations*.
- **1-to-N**: An original sentence is aligned to more than one simplified sentence,
  which is important for learning *sentence splitting* operations.
- **N-to-1**: Several original sentences are aligned to a single simplified sentence,
  teaching TS models how to *merge* complex sentences into a simpler output.

The corpora of aligned complex source and simplified target sentences are not only
used for training TS models, but also for evaluating their performance.

2.1.2.1 Text Simplification Corpora

Most corpora for learning TS transformations are based on SEW,[3] which is a
Wikipedia project consisting of articles that represent a simplified version of the
corresponding traditional EW[4] entries. They are written in Simple English, a variant of the English language that uses a simpler vocabulary and sentence structure.
SEW primarily targets learners of the English language, but it may also aid native-
language children and adults with reading difficulties. Several techniques have been
explored to produce alignments of original sentences from EW and simplified counterparts from SEW, which will be detailed in the following.

PWKP   The first large-scale TS corpus, the PWKP dataset,[5] was gathered by Zhu
et al. (2010). They align complex EW sentences to SEW counterparts over 65K
Wikipedia articles based on their tf-idf cosine similarity, which is measured at sentence level between all sentences of each article pair. Sentences whose similarity
surpasses a certain pre-defined threshold are aligned, resulting in 108K complex-
simple sentence pairs. Both 1-to-1 and 1-to-N alignments are included, with the

---

[3] http://simple.wikipedia.org

[4] http://en.wikipedia.org

[5] https://www.informatik.tu-darmstadt.de/ukp/research_6/data/sentence_simplification/
simple_complex_sentence_pairs/index.en.jsp

latter representing split examples. The test set contains 100 instances, which quickly became the standard benchmark dataset for evaluating TS approaches.

C&K-1 and C&K-2   In Coster and Kauchak (2011b), another Wikipedia-based TS corpus was introduced, C&K-1.[6] Using tf-idf cosine similarity, the authors first align paragraphs of equivalent EW and SEW articles. They then search for the best global sentence alignment following the approach described in Barzilay and Elhadad (2003). It applies a dynamic programming algorithm that takes context into account, based on the idea that the similarity between two sentences is affected by the similarity between pairs of surrounding sentences. In that way, 137K sentences from EW are automatically aligned to corresponding SEW sentences over 10K article pairs, covering 1-to-1, 1-to-2, and 2-to-1 alignments. The dataset is split into 124K instances for training, 12K instances for development and 1.3K instances for testing.

Later, Kauchak (2013) compiled the C&K-2 dataset.[7] Contrary to the previously collected C&K-1 corpus, which was extracted over a small subset of 10K SEW articles, C&K-2 is now derived from more than 60K articles, resulting in 167K aligned sentence pairs.

AlignedWL   Woodsend and Lapata (2011) adopt the process of Coster and Kauchak (2011b) to compile another version of a SEW-EW-based corpus, based on 15K article pairs. The resulting dataset consists of 142K instances of aligned original EW and corresponding simplified SEW sentences.

RevisionWL   Exploiting revision histories in Wikipedia articles, Woodsend and Lapata (2011) propose another approach for compiling a parallel dataset of aligned complex-simple sentence pairs that leverages Wikipedia's collaborative editing process. Whenever an editor changes the content of an article, they need to comment on what the change was and the reason for it. To identify suitable revisions, the authors look for specific keywords in the revision comments of SEW articles, such as "simple", "clarification" or "grammar". Each selected revision is then compared to the previous version. In a first step, modified sections are identified, using the Unix command `diff`. Next, sentences within the sections are aligned through the `dwdiff` Unix command. This process resulted in 15K paired sentences.

More recently, more sophisticated techniques for measuring the similarity between pairs of sentences were proposed for collecting parallel corpora of complex-

---

[6] http://www.cs.pomona.edu/~dkauchak/simplification/

[7] https://cs.pomona.edu/~dkauchak/simplification/

simple sentence pairs for the task of TS, including in particular the EW-SEW dataset (Hwang et al., 2015) and the sscorpus (Kajiwara and Komachi, 2016).

EW-SEW Hwang et al. (2015) introduce a sentence alignment method for TS that is based on a greedy search over the sentences and a word-level similarity measure which is built on top of Wiktionary[8] and incorporates structural similarity represented in syntactic dependencies. By leveraging synonym information and word-definition co-occurrence in Wiktionary, they first create a graph. Then, semantic similarity is measured by counting the number of shared neighbors between words. This word-level similarity metric is then combined with a similarity score that leverages dependency structures. The resulting final similarity rate is used by a greedy algorithm to match a simple sentence to a corresponding original sentence, limiting the generated dataset to 1-to-1 alignments. Based on this approach, the authors align sentences from 22K EW and SEW articles, resulting in a large-scale parallel corpus of 392K automatically aligned complex-simple sentence pairs.

sscorpus Kajiwara and Komachi (2016) propose several sentence similarity measures based on alignments between word embeddings (Song and Roth, 2015; Kusner et al., 2015). For any pair comprising a complex sentence and a simple counterpart, their best metric first aligns each word in the complex sentence with the most similar word in the simple sentence and then averages the similarities for all words in the sentence. All pairs whose similarity exceeds a certain threshold are kept. The authors use this metric to build a parallel corpus for TS from 127K article pairs from EW and SEW, resulting in a dataset of 493K aligned original and simplified sentences.[9]

WikiLarge Through combining several previously created Wikipedia-based TS corpora, Zhang and Lapata (2017) construct WikiLarge.[10] It includes the C&K-2 dataset proposed in Kauchak (2013), a revised version of the PWKP corpus of Zhu et al. (2010) containing 89K sentence pairs ("WikiSmall"), and the aligned and revision sentence pairs from Woodsend and Lapata (2011). The four datasets add up to 300K sentence pairs for training. For testing, the 359 test sentences from PWKP Turk (see below) are used. The WikiLarge corpus has become the most common dataset for training neural Seq2Seq TS approaches.

---

[8] Wiktionary is a free dictionary in the format of a wiki so that everyone can edit its content. It is available under https://en.wiktionary.org.

[9] https://github.com/tmu-nlp/sscorpus

[10] https://github.com/XingxingZhang/dress/tree/master/all-system-output

PWKP Turk With the objective of building a corpus for tuning and testing TS models, Xu et al. (2016) compile the PWKP Turk dataset.[11] It consists of multiple human reference simplifications whose focus is on paraphrasing simplification operations rather than syntactic transformations, thus severely restricting the range of rewriting operations that may be assessed on the basis of this corpus. To create this dataset, the authors select 2,359 sentence pairs from PWKP where the original sentence and the corresponding simplified version are of similar length, assuming that paraphrase-only simplifications are more likely in such a case. The source side of each pair is rewritten by eight Amazon Mechanical Turk workers into a simplified version. In this process, only lexical simplifications are allowed, whereas syntactic simplification operations are prohibited. In the end, the sentences are partitioned into 2,000 pairs for development and 359 for testing. For training, most TS approaches make use of the WikiLarge corpus. To date, PWKP Turk is the most widely used corpus to evaluate and compare TS systems that were trained on Wikipedia data.

ASSET With ASSET (Abstractive Sentence Simplification Evaluation and Tuning),[12] Alva-Manchego et al. (2020) present a generalization of PWKP Turk with a more varied set of rewriting operations. The authors argue that current publicly available evaluation datasets are ill suited to assess the ability of TS models in more realistic settings, since they do not cover the full range of simplification operations, but rather are focused on only one type of transformation. Therefore, they collect simplifications that encompass a richer set of rewrites, including a large number of examples for sentence splitting, paraphrasing and compression. The simplifications are collected via crowdsourcing, using the same set of original sentences as for the PWKP Turk dataset. To obtain a great variability in the simplified instances with regard to the types of transformation operations, the annotators are instructed to apply multiple rewriting transformations for each simplification. The resulting TS corpus consists of 23,590 human simplifications, with 10 simplifications per original sentence.

The TS resources listed above represent distinct versions of parallel corpora of complex source and simplified target sentences that were produced from EW and SEW using different alignment methods. Table 2.15 summarizes some of their features. Below, we point out the characteristics of such Wikipedia-based corpora that qualify [+] or disqualify [-] them for TS research:

---

[11] https://github.com/cocoxu/simplification

[12] https://github.com/facebookresearch/asset

[+] Scale With over 100K instances, the presented datasets provide a large number of aligned pairs of original and simplified sentences for training and testing TS models. The only exception is the RevisionWL corpus, which contains as few as 15K instances, making it by far the smallest of all Wikipedia-based resources.[13] At the other end of the scale are the EW-SEW and sscorpus, as well as WIKI-AUTO (see below), comprising between 400K and 670K instances.

[+] Public availability The corpora of aligned complex source and simplified target sentences that are extracted from corresponding EW and SEW articles are

Table 2.15 Summary of parallel corpora extracted from EW and SEW. An original sentence can be aligned to one (1-to-1) or more (1-to-N) simplified sentences. A (*) indicates that some aligned simplified sentences may not be unique, i.e. they appear more than once in the corpus (Alva-Manchego et al., 2020). A (†) indicates that this corpus is not publicly available. Therefore, we were not able to determine the percentage of split instances contained in it. The scores in brackets denote the percentage of instances of the respective dataset where the input is aligned to more than one simplified sentence, indicating splitting transformations

Corpus	Domain	Instances	Alignment Types
PWKP	Wikipedia	108K	1-to-1, **1-to-N** (6.11%)
C&K-1	Wikipedia	137K	1-to-1, **1-to-2** (16.02%), 2-to-1
RevisionWL	Wikipedia	15K	1-to-1*, **1-to-N*†**, N-to-1*
AlignedWL	Wikipedia	142K	1-to-1, **1-to-N†**
C&K-2	Wikipedia	167K	1-to-1, **1-to-2** (14.52%), 2-to-1
EW-SEW	Wikipedia	392K	1-to-1
sscorpus	Wikipedia	493K	1-to-1
WikiLarge	Wikipedia	286K	1-to-1*, **1-to-N*** (9.64%), N-to-1*
PWKP Turk	Wikipedia	2,359	1-to-1, **1-to-N** (3.68%)
ASSET	Wikipedia	2,359	1-to-1, **1-to-N** (25.16%)
WIKI-AUTO	Wikipedia	488K	1-to-1

---

[13] PWKP Turk and ASSET, too, are small-scale TS corpora. However, they are intended only for tuning and evaluation TS approaches, not for training them.

publicly available, supporting a reproducible cross-system comparison of TS approaches.

[–] **Non-expert authoring of SEW articles** The articles in SEW are not written by linguistic experts, but by volunteers without a well-defined objective in mind. They are supposed to adhere to the Basic English guidelines on how to create syntactically simple sentences, e.g. by avoiding compound sentences and presenting the components of a sentence in subject-verb-object order (Alva-Manchego et al., 2020). However, there is no guarantee that the individual editors follow these guidelines when authoring a SEW article.

[–] **Lack of simplification** Indeed, SEW was found to contain many sentences that are poorly simplified. For instance, a study conducted by Yasseri et al. (2012) has revealed that, even though SEW articles use fewer complex words and shorter sentences, their syntactic complexity is about the same as EW texts, according to their part-of-speech n-gram distribution. These findings were supported by Xu et al. (2015), who examined a random sample of 200 sentences from PWKP, determining that in 33% of the sentence pairs the SEW sentence is not simplified, but presents the same level of complexity as its allocated original counterpart. Even with regard to the pairs where the SEW sentence indeed represents a simplification of its corresponding sentence from EW, the modification is often just minimal: only a few words are simplified, i.e. substituted or deleted, while the remaining part is left unchanged. Furthermore, Coster and Kauchak (2011b) found that in their corpus C&K-1 27% of the aligned complex-simple sentence pairs are identical, which could induce models to learn not to change an original sentence, or to perform very conservative rewriting transformations.

[–] **Noisy data in the form of misalignments** The alignments between complex source and simplified target sentences are automatically produced, making them prone to alignment errors. For example, according to Xu et al. (2015), 17% of the aligned instances from PWKP are misalignments. Moreover, in RevisionWL, an original sentence may be aligned to simplified sentences that appear more than once in the corpus. Alva-Manchego et al. (2020) claim that such a repetition of sentences signifies misalignments. While pairs where the original sentence is identical to its simplified counterpart are important for learning when not to simplify, misalignments add noise to the data and prevent models from learning how to perform the TS transformations accurately (Alva-Manchego et al., 2020).

[–] **Lack of variety of simplification transformations** Most TS corpora do not cover the full range of TS operations. Instead, they typically focus on a single type of transformation. As a result, the instances contained in those datasets offer only a limited variability of simplification rewrites. For instance, PWKP

Turk is restricted to paraphrasing simplifications. Moreover, regarding PWKP, Xu et al. (2015) showed a tendency towards lexical simplification and compression operations, with deletion-only transformations accounting for 21% of the simplifications, paraphrase-only rewrites amounting to 17%, and combinations of deletions and paraphrases to 12% of the simplifications performed on the dataset. In addition, EW-SEW and sscorpus are limited to 1-to-1 instances, hence ruling out the possibility to learn how to split complex sentences. Thus, except for Alva-Manchego et al. (2020), who aim for covering a richer set of rewriting transformations, sentence splitting operations are largely neglected in the Wikipedia-based TS corpora presented above.

Newsela To overcome these problems, Xu et al. (2015) compile a new corpus, Newsela,[14] which is made up of 1,130 news articles that have been manually simplified by professional editors in order to meet the readability requirements for children at different education grade levels. For this purpose, each article has been re-written four times, thus expressing five different levels of complexity in total. As the texts are produced with the intent of being aligned, there is a higher agreement between the simplified and complex news articles in comparison with equivalent EW and SEW articles. Moreover, for each source sentence multiple simplified versions are available. Consequently, automatically aligning Newsela sentences is more straightforward and reliable than for Wikipedia sentences, providing a high quality dataset for building and evaluating TS approaches. Various different techniques for producing sentence alignments were proposed so far, reaching from aligning adjacent article versions only (Štajner and Nisioi, 2018; Scarton et al., 2018; Alva-Manchego et al., 2017) to generating pairs between all versions (Scarton and Specia, 2018).

The range of simplification operations covered in the Newsela corpus are displayed in Figure 2.1, including their proportion therein. As can be seen, the Newsela dataset includes a richer set of rewriting transformations as compared to the Wikipedia-based corpora discussed above, offering more variability in the simplified instances. However, there is still a strong tendency towards compression and lexical simplification transformations, rather than more complex syntactic modifications. Another major drawback of the Newsela corpus is that it is not open access, but only available upon request. Accordingly, researchers are not allowed to produce and publicly share sentence alignments and data splits with the research community, hampering reproducibility.

---

[14] https://newsela.com/data/

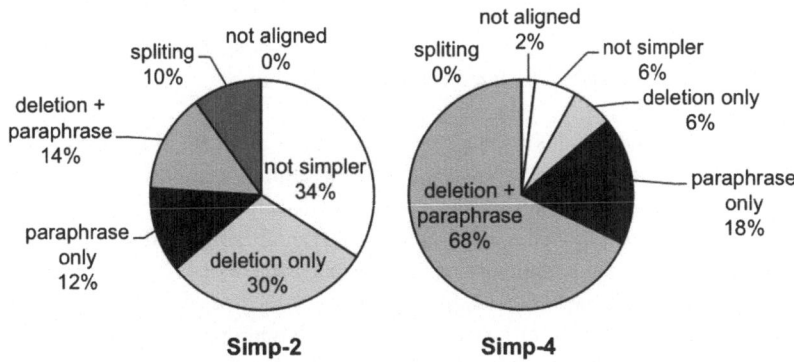

Figure 2.1 Classification of aligned sentence pairs from the Newsela corpus (based on a random sample of 50 sentence pairs) (Xu et al., 2015)

Since its introduction, several works have made use of the Newsela corpus for evaluating the performance of the TS approach they propose, often along with PWKP in order to demonstrate domain-independence (e.g., Zhang and Lapata (2017), Vu et al. (2018), Guo et al. (2018)).

Newsela-Auto and Wiki-Auto To improve the reliability of the sentence alignments in both Newsela- and Wikipedia-based TS datasets, Jiang et al. (2020) propose a new efficient annotation methodology for creating parallel corpora of complex source and simplified target sentences. It is based on the idea to first manually align a small set of sample sentences, which are then used to train a neural CRF sentence aligner that fine-tunes BERT (Devlin et al., 2019) to capture the semantic similarity between two sentences and leverages the sequential nature of sentences in parallel documents. Finally, the remaining sentences are automatically aligned using the trained alignment model. In this way, 1,882 articles from Newsela are aligned, resulting in the largest TS corpus to date, comprising 666K pairs of complex source and simplified target sentences ("NEWSELA- AUTO"). Moreover, the same method is applied to 138K pairs of articles from EW and SEW, yielding a TS dataset of 488K aligned complex-simple sentence pairs ("WIKI- AUTO").[15]

2.1.2.2 Sentence Splitting Corpora
As pointed out in the previous section, existing TS corpora contain only a small number of split examples, thus largely ignoring sentence splitting transformations.

---

[15] https://github.com/chaojiang06/wiki-auto

Instead, they show a tendency towards lexical simplification and compression operations. Accordingly, these datasets are ill suited for learning how to split sentences into shorter, syntactically simplified components.

WebSplit   To address this shortcoming, Narayan et al. (2017) introduce a new TS task, Split-and-Rephrase. Its goal is to split structurally complex sentences into several shorter ones, while making the rephrasings that are necessary to preserve the meaning of the input, as well as the grammaticality of the simplified sentences. For training and testing models attempting to solve this task, they compile the WebSplit corpus,[16] the first TS dataset that explicitly addresses the task of sentence splitting, while abstracting away from lexical and deletion-based operations. It is derived in a semi-automatic way from the WebNLG corpus (Gardent et al., 2017), which contains 13K pairs of items, with each of them consisting of a set of RDF triples and one or more texts verbalising them. The resulting WebSplit dataset is composed of over one million entries, with each of them containing (1) a MR of an original sentence in the form of a set of RDF triples ⟨*subject*; *predicate*; *object*⟩; (2) the original sentence to which the MR corresponds; and (3) several pairs of MR and associated sentence, representing valid splits (i.e. simple sentences) of the original sentence.

   Aharoni and Goldberg (2018) criticize the data split proposed by Narayan et al. (2017). They observed that 99% of the simple sentences (which make up for more than 89% of the unique ones) contained in the validation and test sets also appear in the training set. Consequently, instead of learning how to split and rephrase complex sentences, models that are trained on this dataset will be prone to learn to memorize entity-fact pairs. Hence, this split is not suitable for measuring a model's ability to generalize to unseen input sentences. To fix this issue, Aharoni and Goldberg (2018) present a new train-development-test data split[17] where no simple sentence that is contained in the development or test set occurs verbatim in the training set.

Table 2.16 Characteristic example source sentences of the WebSplit corpus

(1)	A Loyal Character Dancer was published by Soho Press, in the United States, where some Native Americans live.
(2)	Dead Man's Plack is in England and one of the ethnic groups found in England is the British Arabs.

---

[16] https://github.com/shashiongithub/Split-and-Rephrase
[17] https://github.com/roeeaharoni/sprp-acl2018

WikiSplit Botha et al. (2018) claim that the sentences from the WebSplit corpus contain fairly unnatural linguistic expressions over only a small vocabulary and a rather uniform sentence structure (see Table 2.16). To overcome these limitations, they present WikiSplit,[18] a sentence splitting dataset of one million sentence pairs that was mined from Wikipedia edit histories. In this dataset, each original sentence is aligned to exactly two simpler ones. To identify edits involving sentences being split and align them, a simple heuristic was applied: adjacent snapshots were compared, searching for pairs of sentences where the trigram prefix and trigram suffix of the original sentence matches, respectively, the trigram prefix of the first simple sentence and the trigram suffix of the second simple sentence. To filter out misaligned pairs, a high-precision BLEU-based heuristic was used, discarding pairs of sentences whose BLEU score falls below an empirically chosen threshold. The resulting dataset provides a rich and varied vocabulary over naturally expressed source sentences that show a diverse linguistic structure, and their extracted splits (see Table 2.17 for some example instances). However, there is only a single split per source sentence in the training set. Accordingly, when a TS model is trained on this dataset, it is susceptible to exhibiting a strong conservatism, splitting each input sentence into exactly two output sentences only. Consequently, the resulting simplified sentences are often still comparatively long and complex, mixing multiple, potentially semantically unrelated propositions that may be difficult to analyze in downstream Open IE tasks.

Table 2.17 Pairs of aligned complex source and simplified target sentences of WikiSplit

Complex source	Starring Meryl Streep, Bruce Willis, Goldie Hawn and Isabella Rossellini, the film focuses on a childish pair of rivals who drink a magic potion that promises eternal youth.
Simplified output	Starring Meryl Streep, Bruce Willis, Goldie Hawn and Isabella Rossellini. The film focuses on a childish pair of rivals who drink a magic potion that promises eternal youth.
Complex source	The Assistant Attorney in Orlando investigated the modeling company, and decided that they were not doing anything wrong, and after Pearlman's bankruptcy, the company emerged unscathed and was sold to a Canadian company.
Simplified output	The Assistant Attorney in Orlando investigated the modeling company, and decided that they were not doing anything wrong. After Pearlman's bankruptcy, the modeling company emerged unscathed and was sold to a Canadian company.

---

[18] https://github.com/google-research-datasets/wiki-split

HSplit   Sulem et al. (2018a) present HSplit,[19] a small-scale human-generated gold standard sentence splitting corpus that was built upon the 359 test sentences from PWKP Turk. It is a multi-reference dataset that was specifically designed for assessing the sentence splitting operation. While the simplified references from PWKP Turk do not contain examples for sentence splitting or content deletion, but rather emphasize lexical simplifications, HSplit focuses on structural simplification operations, with the aim of providing a high quality dataset for the evaluation of systems that focus on the task of sentence splitting. In HSplit, each complex source sentence is modified by two annotators, according to two sets of sentence splitting guidelines. In the first setting, the annotators are instructed to split a given source sentence as much as possible, while preserving the grammaticality and meaning of the input, whereas in the second setting, the annotators are encouraged to split only in cases where it explicitly simplifies the original sentence. Since the simplification references were manually created, the HSplit corpus is of high quality. However, due to its small scale, it is not well suited for training sentence splitting models, but rather for evaluating their performance.

Table 2.18 Summary of the main properties of the above sentence splitting corpora. The scores in brackets signify the percentage of instances of the respective dataset where the input is aligned to more than one simplified sentence, indicating splitting transformations

Corpus	Domain	Instances	Alignment Types
WebSplit	Wikipedia	1,040,430	**1-to-N** (100.00%)
WebSplit-v2	Wikipedia	1,040,430	**1-to-N** (100.00%)
WikiSplit	Wikipedia	1,004,944	**1-to-2** (100.00%)
HSplit	Wikipedia	359	1-to-1, **1-to-N** (73.47%)[a]

[a] Average score calculated over the four simplified references. With 64.90%, *HSplit-3* is the reference with the lowest amount of split instances, while *HSplit-2* contains by far the largest number of split examples (86.91%).

Table 2.18 summarizes the main properties of the sentence splitting corpora described above.

2.1.2.3 Summary
As pointed out in Tables 2.15 and 2.18, there is a lack of large-scale TS corpora that allow to train TS models how to split complex sentences into a fine-grained simplified output. Focusing on paraphrasing and deletion operations, existing TS datasets contain only a small proportion of split instances of up to 16%, if any.

---

[19] https://github.com/eliorsulem/HSplit-corpus

The only exception is the very recent ASSET corpus, which at least includes about 25% of examples where the input is aligned to more than one simplified sentence, suggesting splitting transformations.

To overcome this limitation, researchers recently started to collect datasets that explicitly address the task of sentence splitting. However, each of them has its own shortcomings, ranging from a size which is too small to be suitable for use in data-driven approaches to insufficient simplifications of the reference sentences and a lack of syntactic variety in the example instances.

2.2 Text Coherence

When simplifying the structure of sentences without considering discourse implications, the rewriting may easily result in a loss of cohesion, hindering the process of making sense of the text (Siddharthan, 2002; Siddharthan, 2006). Thus, preserving the coherence of the output is crucial to maintain its interpretability in downstream applications.

2.2.1 Discourse-level Syntactic Simplification

The vast majority of existing structural TS approaches do not take into account discourse level effects that arise from splitting long and syntactically complex sentences into a sequence of shorter utterances (Alva-Manchego et al., 2020). However, not considering the semantic context of the output propositions, the simplified text is prone to lack coherence, resulting in a set of incoherent propositions that miss important contextual information that is needed to infer the true meaning of the simplified output. Though, two notable exceptions have to be mentioned.

Siddharthan (2006) was the first to use discourse-aware cues in the simplification process, with the goal of generating a coherent output that is accessible to a wider audience. For this purpose, he adds a regeneration stage after the transformation process where, amongst others, simplified sentences are reordered, appropriate determiners are chosen (see Table 2.19) and cue words in the form of rhetorical relations are introduced to connect them, thus making information easier to comprehend for people with reduced literacy. Though, as opposed to our approach, where a semantic relationship is established for each output sentence, only a comparatively low number of sentences is linked by such cue words in Siddharthan (2006)'s framework.

Table 2.19 Adequate determiner choice to maintain cohesion (Siddharthan, 2006)

Complex source	The man who had brought it in for an estimate returned to collect it.
Cohesive simplified output	A man had brought it in for an estimate. *This* man returned to collect it.

Another approach that simplifies texts on the discourse level was proposed by Štajner and Glavaš (2017) (see Section 2.1.1.1). It makes use of an event extraction system (Glavaš and Štajner, 2015) which operates on the level of discourse to perform content reduction both within a single sentence and within a whole document. For this purpose, a semantically motivated method is applied, eliminating irrelevant information from the input by maintaining only those parts that belong to factual event mentions. Our approach, on the contrary, aims to preserve all the information contained in the source, as illustrated in Table 2.4. However, by distinguishing core from contextual information, we are still able to extract only the key information given in the input.

2.2.2   Discourse Parsing

The task of Discourse Parsing aims to identify discourse relations that hold between textual units in a document (Marcu, 1997). In that way, coherence structures in texts are uncovered. Two major directions in this area were established on the basis of the annotated corpora of the Penn Discourse TreeBank (PDTB) (Miltsakaki et al., 2004) and the RST-DT (Carlson et al., 2002). Whereas PDTB provides a more shallow specification of discourse structures, RST-DT analyzes the hierarchical structure of discourse in texts based on RST (Mann and Thompson, 1988).

In this work, we will focus on the taxonomy of RST, which was introduced by Mann and Thompson (1988). It is one of the most established theories of text organization, characterizing the structure of a text in terms of relations that hold between its spans. It offers a way of reasoning about textual coherence that is based on the rationale that texts commonly present a hierarchical, connected arrangement of their components in which every part plays a specific role, relative to the others (Taboada and Mann, 2006). Accordingly, a text is regarded as fully connected if every span has a particular purpose and is linked with the rest by means of some relation. In RST, two main types of relations are distinguished: nucleus-satellite and multinuclear. The former reflects hypotactic syntactic structures, where the nucleus (N) embodies the central piece of information, while the associated satellite (S) provides secondary information that further specifies its respective nucleus (Mann and Thompson, 1988). Multinuclear relations, in contrast, mirror a paratactic syntax where both spans are considered equally important.

RST has been designed to enable the analysis of texts by making available a set of rhetorical relations to annotate a given text (Mann and Thompson, 1988). Hence, when examining a text, the annotator allocates a specific role to each text span, thereby constructing a rhetorical structure tree with one top-level relation that encompasses the remaining relations at lower levels, thus including every part of the text in one connected whole (Taboada and Mann, 2006). In that way, RST identifies rhetorical relations that hold between units of a text. For instance, the *Evidence* relationship is defined as follows (Table 2.20):

Table 2.20 Definition of the *Evidence* relation (Mann and Thompson, 1988)

Relation name	EVIDENCE
Constraints on N	The reader might not believe N to a degree satisfactory to the writer.
Constraints on S	The reader believes S or will find it credible.
Constraints on the N + S combination	The reader's comprehension of S increases the reader's belief of N.
The effect	The reader's belief of N is increased.

An application of the *Evidence* relation is illustrated in the following example:

[The truth is that the pressure to smoke in junior high is greater than it will be any other time of one's life:]$_N$ [we know that 3,000 teens start smoking each day.]$_S$ (Marcu, 1998)

Here, statistical evidence is provided in the satellite span in order to increase the reader's belief in the claim expressed in the preceding unit (nucleus span).

Multiple different lists of rhetorical relations exist by now. While the original set presented in Mann and Thompson (1988) consists of 25 relations, more recent work has extended this collection. The annotations of the RST-DT corpus, for instance, encompass 53 mono-nuclear and 25 multi-nuclear relations.

As noted in Cetto (2017), approaches to identify rhetorical structure arrangements in texts range from early rule-based approaches (Marcu, 2000) to supervised data-driven models that are trained on annotated corpora such as the RST-DT (Feng and Hirst, 2014; Li et al., 2014). Most commonly, these approaches separate the task of extracting discourse structures into two major subtasks:

(1) elementary discourse unit (EDU) segmentation, where the input text is parti-
     tioned into basic textual units, and
(2) a tree-building step, where the identified EDUs are connected by rhetorical
     relations to form a rhetorical structure tree.

Due to the simplicity of the first task and the high accuracy achieved on it (90% by
Fisher and Roark (2007)), research currently concentrates on the more challenging
task of constructing the tree of rhetorical structures (Feng and Hirst, 2014; Joty et al.,
2015). Recent approaches often solve this task by training two individual classifiers:
a binary structure classifier for determining pairs of consecutive text spans that are to
be joined together, and a multi-class relation classifier for identifying the rhetorical
relation that holds between them (Hernault et al., 2010; Li et al., 2014; Feng and
Hirst, 2014; Joty et al., 2015). The most prominent features that are used by state-of-
the-art systems for this step are based on lexical and syntactic information, including
cue phrases, part-of-speech tags, and parse tree structures (Feng and Hirst, 2014; Li
et al., 2014; Liu and Lapata, 2017; Lin et al., 2019).

     Although the segmentation step is related to the task of sentence splitting, it
is not possible to simply use an RST parser for the sentence splitting subtask in

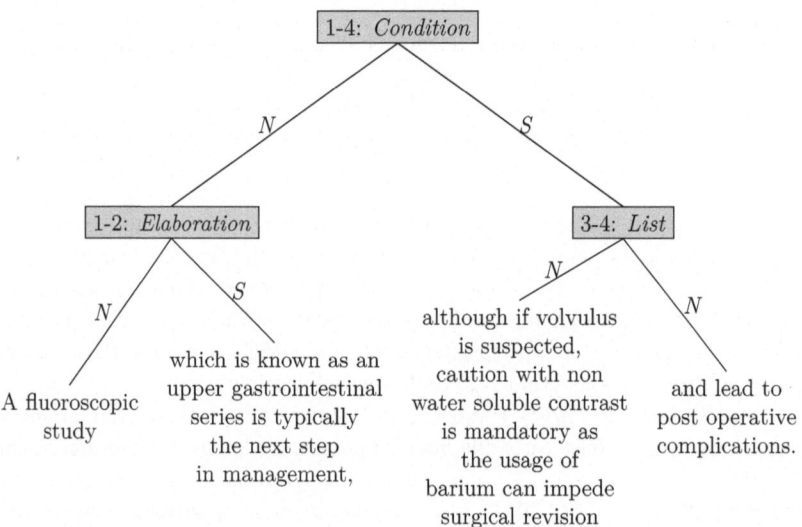

Figure 2.2 Rhetorical structure tree of the example sentence from Table 1.1 generated using
the RST parser proposed by tjispseisensteinsps2014spsrepresentation (*N:* nucleus, *S:* satellite)

our TS approach. As illustrated in Figure 2.2, such a parser does not return grammatically sound sentences, but rather just copies the components that belong to the corresponding EDUs from the source. In order to reconstruct proper sentences, rephrasing is required. For this purpose, amongst others, referring expressions have to be identified, and phrases have to be rearranged and inflected. Moreover, the textual units resulting from the segmentation process are too coarse-grained for our purpose, since RST parsers mostly operate on clausal level. The goal of our approach, though, is to split the input into minimal semantic units, which requires to go down to the phrasal level in order to produce a much more fine-grained output in the form of minimal propositions.

2.3   Open Information Extraction

IE turns the unstructured information expressed in natural language (NL) text into a structured representation (Jurafsky and Martin, 2009) in the form of relational tuples consisting of a set of arguments and a phrase denoting a semantic relation between them: $\langle arg1; rel; arg2 \rangle$. Traditional approaches to IE focus on answering narrow, well-defined requests over a predefined set of target relations on small, homogeneous corpora. To do so, they take as input the target relation along with hand-crafted extraction patterns or patterns learned from hand-labeled training examples (e.g., Agichtein and Gravano (2000), Brin (1998), Riloff and Jones (1999)). Consequently, shifting to a new domain requires the user to not only name the target relations, but also to manually define new extraction rules or to annotate new training data by hand. Thus, those systems rely on extensive human involvement. In order to reduce the manual effort required by IE approaches, Banko et al. (2007) introduced a new extraction paradigm: Open IE. Unlike traditional IE methods, Open IE is not limited to a small set of target relations known in advance, but rather extracts all types of relations found in a text. In that way, it facilitates the domain-independent discovery of relations extracted from text and scales to large, heterogeneous corpora such as the Web. Hence, Banko et al. (2007) identified three major challenges for Open IE systems:

Automation   Open IE systems must *rely on unsupervised extraction strategies*, i.e. instead of specifying target relations in advance, possible relations of interest must be automatically detected while making only a single pass over the corpus. Moreover, the manual labor of creating suitable training data or extraction patterns must be reduced to a minimum by requiring only a small set of hand-tagged seed instances or a few manually defined extraction patterns.

Corpus Heterogeneity Heterogeneous datasets form an obstacle for tools that perform a deep linguistic analysis, such as syntactic or dependency parsers, since they commonly work well when trained on and applied to a specific domain, but are prone to produce incorrect results when used in a different genre of text. Furthermore, Named Entity Recognition (NER) is unsuitable to target the variety and complexity of entity types on the Web. As Open IE systems are intended for *domain-independent usage*, such tools should be avoided in favor of shallow parsing methods such as part-of-speech taggers.

Efficiency In order to *readily scale to large amounts of text*, Open IE systems must be computationally efficient. Enabling fast extraction over huge datasets, shallow linguistic features, like part-of-speech tags, are to be preferred over deep features that are derived from parse trees.

These criteria were first implemented in the Open IE system TEXTRUNNER, which was presented together with the task definition in Banko et al. (2007). This seminal work triggered a lot of research effort in this area, resulting in a multitude of proposed approaches that often do not strictly adhere to these initial guidelines. For example, to date, Open IE systems are commonly evaluated on rather small-scale, domain-dependent corpora. In addition, recent approaches frequently rely on the output of a dependency parser to identify extraction patterns, thereby hurting the domain-independence and efficiency assumptions.

2.3.1 Approaches to Open Information Extraction

A large body of work on the task of Open IE has been described since its introduction by Banko et al. (2007). Existing Open IE approaches make use of a set of patterns in order to extract relational tuples from a sentence, each consisting of a set of arguments and a phrase that expresses a semantic relation between them. Such extraction patterns are either hand-crafted or learned from automatically labeled training data, as shown below.

2.3.1.1 Early Approaches based on Hand-crafted Extraction Patterns and Self-supervised Learning

Early approaches to the task of Open IE are either based on hand-crafted extraction patterns, where a human expert manually defines a set of extraction rules, or self-supervised learning, where the system automatically finds and labels its own training examples.

Binary Relations   The line of work on Open IE begins with TEXTRUNNER (Banko et al., 2007), a self-supervised learning approach that consists of three modules. First, given a small sample of sentences from the Penn Treebank (Marcus et al., 1993), the learner applies a dependency parser to heuristically identify and label a set of extractions as positive and negative training examples. This data is then used as input to a Naive Bayes classifier which learns a model of trustworthy relations using unlexicalized part-of-speech and noun phrase chunk features. The self-supervised nature mitigates the need for hand-labeled training data, and unlexicalized features help scale to the multitudes of relations found on the Web. The second component, the extractor, then generates candidate tuples from unseen text by first identifying pairs of noun phrase arguments and then heuristically designating each word in between as part of a relational phrase or not. Next, each candidate extraction is presented to the classifier, which keeps only those labeled as trustworthy. The restriction to the use of shallow features in this step makes TEXTRUNNER highly efficient. Finally, a redundancy-based assessor assigns a probability to each retained tuple based on the number of sentences from which each extraction was found, thus exploiting the redundancy of information in Web text and assigning higher confidence to extractions that occur multiple times.

WOE (Wu and Weld, 2010) also learns an open information extractor without direct supervision. It makes use of Wikipedia as a source of training data by bootstrapping from entries in Wikipedia infoboxes, i.e. by heuristically matching infobox attribute-value pairs with corresponding sentences in the article. This data is then used to learn extraction patterns on both part-of-speech tags (WOE^{pos}) and dependency parses (WOE^{parse}). Former extractor utilizes a linear-chain CRF to train a model of relations on shallow features which outputs certain text between two noun phrases when it denotes a relation. Latter approach, in contrast, makes use of dependency trees to build a classifier that decides whether the shortest dependency path between two noun phrases indicates a semantic relation. By operating over dependency parses, even long-range dependencies can be captured. Accordingly, when comparing their two approaches, Wu and Weld (2010) show that the use of dependency features results in an increase in precision and recall over shallow linguistic features, though, at the cost of extraction speed, hence negatively affecting the scalability of the system.

Fader et al. (2011) propose REVERB, a shallow extractor that addresses three common errors of previously presented Open IE approaches: the output of such systems frequently contains a great many of *uninformative extractions* (i.e., extractions that omit critical information), *incoherent extractions* (i.e., extractions where the relational phrase has no meaningful interpretation) and *overly-specific relations* that convey too much information to be useful in downstream semantic tasks.

$$V \mid VP \mid VW^*P$$
$$V = \text{verb particle? adv?}$$
$$W = (\text{noun} \mid \text{adj} \mid \text{adv} \mid \text{pron} \mid \text{det})$$
$$P = (\text{prep} \mid \text{particle} \mid \text{inf. marker})$$

Figure 2.3 REVERB's part-of-speech-based regular expression for reducing incoherent and uninformative extractions

REVERB improves over those approaches by introducing a syntactic constraint that is expressed in terms of simple part-of-speech-based regular expressions (see Figure 2.3), covering about 85% of verb-based relational phrases in English texts, as a linguistic analysis has revealed. In that way, the amount of incoherent and uninformative extractions is reduced. Moreover, in order to avoid overspecified relational phrases, a lexical constraint is presented which is based on the idea that a valid relational phrase should take many distinct arguments in a large corpus. Besides, while formerly proposed approaches start with the identification of argument pairs, REVERB follows a relation-centric method by first determining relational phrases that satisfy above-mentioned constraints, and then finding a pair of noun phrase arguments for each such phrase.

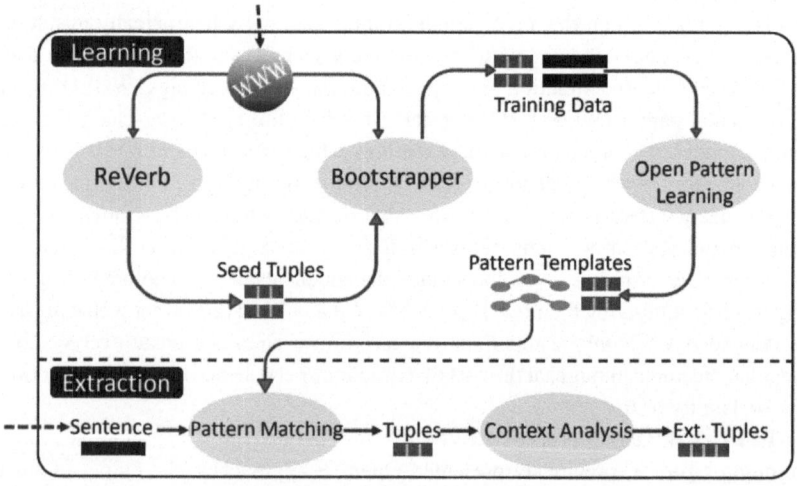

Figure 2.4 OLLIE's system architecture (Mausam et al., 2012). OLLIE begins with seed tuples from REVERB, uses them to build a bootstrap learning set, and learns open pattern templates, which are applied to individual sentences at extraction time

OLLIE (Mausam et al., 2012) follows the idea of bootstrap learning of patterns based on dependency parse paths. However, while WOE relies on Wikipedia-based bootstrapping, OLLIE applies a set of high precision seed tuples from its predecessor system REVERB to bootstrap a large training set over which it learns a set of extraction pattern templates using dependency parses (see Figure 2.4). In contrast to previously presented systems that fully ignore the context of a tuple and thus extract propositions that are not asserted as factual, but are only hypothetical or conditionally true, OLLIE includes a context-analysis step in which *contextual information* from the input sentence around an extraction is analyzed to expand the output representation by adding attribution and clausal modifiers, if necessary, and thus increasing the precision of the system (for details, see Section 2.3.1.3). Moreover, OLLIE is the first Open IE approach to identify not only verb-based relations, but also relationships mediated by nouns and adjectives (see OLLIE's extractions in Table 2.21). In that way, it *expands the syntactic scope of relational phrases* to cover a wider range of relation expressions, resulting in a much higher yield (at comparable precision) as compared to previous systems.

Table 2.21 illustrates by means of an example sentence how the proposed Open IE systems evolve over time by addressing particular issues that were identified in prior approaches.

Table 2.21 Evolution of early Open IE approaches illustrated by means of the relations extracted from the following input sentence: *"The novelist Franz Kafka is the author of a short story entitled 'The Metamorphosis'."*

Approach	Extracted Tuple	Pros	Cons
TEXTRUNNER	⟨*The novelist Franz Kafka*; *is*; *the author of a short story*⟩		uninformative relational phrases
REVERB	⟨*The novelist Franz Kafka*; *is the author of*; *a short story*⟩	extraction of coherent and meaningful relational phrases	limitation to verb-mediated relations
OLLIE	⟨*Franz Kafka*; *is the author of*; *a short story*⟩, ⟨*Franz Kafka*; *is*; *a novelist*⟩, ⟨*a short story*; *be entitled*; *'The Metamorphosis'*⟩	wider syntactic scope of relational phrases	still limited to binary relations

N-ary Relations The Open IE systems presented above focus on the extraction of binary relations, which commonly leads to extraction errors such as incomplete, uninformative or erroneous tuples. To address this problem, Akbik and Löser (2012) present KRAKEN. It is the first approach that is specifically built for capturing complete facts from sentences by *gathering the full set of arguments for each relational phrase* within a sentence, thus producing tuples of arbitrary arity. The identification of relational phrases and their corresponding arguments is based on hand-written extraction rules over typed dependency parses.

EXEMPLAR (Mesquita et al., 2013) applies a similar approach for extracting n-ary relations, using hand-crafted patterns based on dependency parse trees to detect a relation trigger and the arguments connected to it. Based on the task of SRL, whose key idea is to classify semantic constituents into different semantic roles (Christensen et al., 2010) (see Section 2.4), it assigns each argument its corresponding role (such as subject, direct object or prepositional object).

Table 2.22 From binary to n-ary relations. The extractions above compare REVERB's binary relational tuple with EXEMPLAR's n-ary relational tuple when given the following input sentence: *"Franz Kafka was born into a Jewish family in Prague in 1883."*, showing that the latter is much more informative and complete

Approach	Extracted Tuple	Pros	Cons
REVERB	⟨*Franz Kafka*; *was born into*; *a Jewish family*⟩		limitation to binary relations
EXEMPLAR	⟨*Franz Kafka*; *was born*; *(into) a Jewish family*; *(in) Prague*; *(in) 1883*⟩	extraction of complete facts by gathering the full set of arguments for a relational phrase (n-ary relations)	prone to erroneous extractions on syntactically complex sentences

Table 2.22 depicts the added value of extracting n-ary relations that capture the full set of arguments for a relational phrase, thus generating informative extractions that represent complete facts from the input. In contrast, Open IE approaches that are limited to the extraction of binary relations are prone to missing some important aspects of a fact or event stated in the source sentence.

Table 2.23 summarizes the properties of early approaches to the task of Open IE, taking into account the features that they use to specify or learn the extraction patterns, the arity of the relations that they capture and the syntactic scope of relational phrases that they cover.

Table 2.23 Properties of early approaches to Open IE

Approach	Features	Arity	Type of Relations
TEXTRUNNER	shallow syntax (part-of-speech tagging and noun-phrase chunking)	binary	verb-based
WOE	shallow syntax (part-of-speech tagging for WOE^{pos}) and dependency parse (for WOE^{parse})	binary	verb-based
OLLIE	dependency parse	binary	mediated by verbs, nouns and adjectives
REVERB	shallow syntax (part-of-speech tagging and noun phrase chunking)	binary	verb-based
KRAKEN	dependency parse	n-ary	verb-based
EXEMPLAR	dependency parse	binary and n-ary	verb- and noun-based

2.3.1.2 Paraphrase-based Approaches

Noticing that previously proposed Open IE approaches are *prone to erroneous extractions on syntactically complex sentences*, more recent work is based on the idea of incorporating a *sentence re-structuring stage*. Its goal is to transform complex sentences, where relations are spread over several clauses or presented in a non-canonical form, into a set of syntactically simplified independent clauses that are easy to segment into Open IE tuples. Thus, they aim to improve the coverage and accuracy of the extractions. An example of such a paraphrase-based Open IE approach is ClausIE (Del Corro and Gemulla, 2013), which exploits linguistic knowledge about the grammar of the English language to map the dependency relations of an input sentence to clause constituents. In that way, a set of coherent clauses presenting a simple linguistic structure is derived from the input. Then, the type of each clause is determined by combining knowledge of properties of verbs (with the help of domain-independent lexica) with knowledge about the structure of input clauses. Finally, based on its type, one or more tuples are extracted from each clause, each representing different pieces of information. The basic set of patterns used for this task is shown in Table 2.24.

In the same vein, Schmidek and Barbosa (2014) propose a strategy to break down structurally complex sentences into simpler ones by decomposing the original sentence into its basic building blocks via chunking. The dependencies of each two chunks are then determined (one of "connected", "disconnected" or "dependent") using either manually defined rules over dependency paths between words in different chunks or a Naive Bayes classifier trained on shallow features, such as part-of-speech tags and the distance between chunks. Depending on their relationships, chunks are combined into simplified sentences, upon which the extraction process is carried out.

Angeli et al. (2015) present Stanford Open IE, an approach in which a classifier is learned for splitting a sentence into a set of logically entailed shorter utterances by recursively traversing its dependency tree and predicting at each step whether an edge should yield an independent clause or not. In order to increase the usefulness of the extracted tuples for downstream applications, each self-contained clause is then maximally shortened by running natural logic inference (Van Benthem, 2008) over it. In the end, a small set of 14 hand-crafted patterns are used to extract a predicate-argument triple from each utterance. An illustration of this approach is depicted in Figure 2.5.

Arguing that it is hard to read out from a dependency parse the complete structure of a sentence's propositions, since, amongst others, different predications are represented in a non-uniform manner and proposition boundaries are not easy to detect,

Table 2.24 ClausIE's basic patterns for relation extraction (Del Corro and Gemulla, 2013). S: Subject, V: Verb, C: Complement, O: Direct object, A: Adverbial, V_i: Intransitive verb, V_c: Copular verb, V_e: Extended-copular verb, V_{mt}: Monotransitive verb, V_{dt}: Ditransitive verb, V_{ct}: Complex-transitive verb

	Pattern	Clause type	Example	Derived clauses
S_1 :	SV_i	SV	AE died.	(AE, died)
S_2 :	SV_eA	SVA	AE remained in Princeton.	(AE, remained, in Princeton)
S_3 :	SV_cC	SVC	AE is smart.	(AE, is, smart)
S_4 :	$SV_{mt}O$	SVO	AE has won the Nobel Prize.	(AE, has won, the Nobel Prize)
S_5 :	$SV_{dt}O_iO$	SVOO	RSAS gave AE the Nobel Prize.	(RSAS, gave, AE, the Nobel Prize)
S_6 :	$SV_{ct}OA$	SVOA	The doorman showed AE to his office.	(The doorman, showed, AE, to his office)
S_7 :	$SV_{ct}OC$	SVOC	AE declared the meeting open.	(AE, declared, the meeting, open)

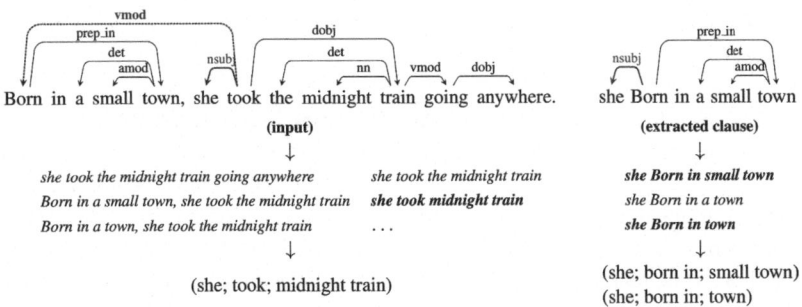

Figure 2.5 An illustration of Stanford Open IE's approach. From left to right, a sentence yields a number of independent clauses. From top to bottom, each clause produces a set of entailed shorter utterances, and segments the ones which match an atomic pattern into a relational tuple (Angeli et al., 2015)

Stanovsky et al. (2016) suggest PROPS. They introduce a semantically-oriented sentence representation that is generated by transforming a dependency parse tree into a directed graph which is tailored to directly represent the proposition structure of an input sentence. Consequently, extracting relational tuples from this novel output format is straightforward. The conversion of the dependency tree into the proposition structure is carried out by a rule-based converter.

Table 2.25 Properties of paraphrase-based approaches to Open IE

Approach	Features	Arity	Type of Relations
ClausIE	dependency parse	binary and n-ary	verb-based relations and noun-mediated relations from appositions and possessives
Schmidek and Barbosa (2014)	two modes: • shallow syntax (part-of-speech tags) • dependency parse	binary and n-ary	verb- and noun-based
Stanford Open IE	dependency parse	binary	verb- and noun-based
PROPS	dependency parse	n-ary	mediated by verbs, nouns and adjectives

Table 2.25 gives an overview of the properties of the paraphrase-based Open IE approaches described above.

2.3.1.3 Systems that Capture Inter-Proposition Relationships

Aforementioned Open IE systems *lack the expressiveness needed for a proper interpretation of complex assertions*, since they ignore the context under which a proposition is complete and correct. Thus, they do not distinguish between information asserted in a sentence and information that is only hypothetical or conditionally true. For example, extracting the relational tuple ⟨*the earth*; *is the center of*; *the universe*⟩ from the sentence *"Early scientists believed that the earth is the center of the universe."* would be inappropriate, since the input is not asserting this proposition, but only noting that it was believed by early scientists (Mausam, 2016). To properly handle such cases, OLLIE introduces an attribution context to indicate that an extracted relational tuple is reported or claimed by some entity:

> (⟨*the earth*; *be the center of*; *the universe*⟩;
> `AttributedTo` *believe*; *Early astronomers*)

In that way, it extends the default Open IE representation of ⟨arg_1; rel; arg_2⟩ tuples with an extra field. Beyond attribution relationships, OLLIE targets the identification of clausal modifiers, such as:

> (⟨*Romney*; *will be elected*; *President*⟩;
> `ClausalModifier` *if*; *he wins five key states*)

Both types of modifiers are identified by matching patterns with the dependency parse of the input sentence. Clausal modifiers are determined by an adverbial clause edge and filtered lexically (the first word of the clause must match a list of cue terms, e.g. *if*, *when*, or *although*), while attribution modifiers are identified by a clausal complement edge whose context verb must match one of the terms given in VerbNet's list of common verbs of communication and cognition (Mausam et al., 2012).

A similar type of output is produced by OLLIE's successor OPENIE- 4 (Mausam, 2016), which combines SRLIE (Christensen et al., 2010) and RELNOUN (Pal and Mausam, 2016). The former is a system that converts the output of an SRL system into an Open IE extraction by treating the verb as the relational phrase, while taking its role-labeled arguments as the corresponding Open IE argument phrases. The latter, in contrast, represents a rule-based Open IE system that extracts noun-mediated relations, thereby paying special attention to demonyms and compound relational nouns. In addition, OPENIE- 4 marks up temporal and spatial arguments by assigning them a T or S label, respectively. For instance, it extracts the following tuple from the example sentence from Table 2.22: $\langle$ *Franz Kafka was born into a Jewish family;* **L:** *in Prague;* **T:** *in 1883* $\rangle$. Lately, its successor OPENIE- 5 was released. It integrates BONIE (Saha et al., 2017) and CALMIE (Saha and Mausam, 2018). While the former focuses on extracting tuples where one of the arguments is a number or a quantity-unit phrase, the latter targets the extraction of relational tuples from conjunctive sentences.

Similar to OLLIE, Bast and Haussmann (2013), who explore the use of contextual sentence decomposition (CSD) for Open IE, advocate to further specify relational tuples with information on which they depend. Their system CSD-IE is based on the idea of paraphrasing-based approaches described in Section 2.3.1.2. Using a set of hand-crafted rules over the output of a constituent parser, a sentence is first split into sub-sequences that semantically belong together, forming so-called "contexts". Each such context now contains a separate fact, yet it is often dependent on surrounding contexts. In order to preserve such inter-proposition relationships, tuples may contain untyped references to other extractions. However, as opposed to OLLIE, where additional contextual modifiers are directly assigned to the corresponding relational tuples, Bast and Haussmann (2013) represent contextual information in the form of separate, linked tuples. To do so, each extraction is given a unique identifier that can be used in the argument slot of an extraction for a later substitution with the corresponding fact by a downstream application. An example for an attribution is shown below (Bast and Haussmann, 2013):

#1: $\langle$*The Embassy*; *said*; *that #2*$\rangle$
#2: $\langle$*6,700 Americans*; *were*; *in Pakistan.*$\rangle$

Another current approach that captures inter-proposition relationships is proposed by Bhutani et al. (2016), who present a nested representation for Open IE that is able to capture high-level dependencies, allowing for a more accurate representation of the meaning of an input sentence. Their system NESTIE uses bootstrapping over a dataset for textual entailment to learn both binary and nested triple representations for n-ary relations over dependency parse trees. These patterns can take on the form of binary tuples $\langle arg_1; rel; arg_2 \rangle$ or nested tuples, such as $\langle \langle arg_1; rel; arg_2 \rangle; rel_2; arg_3 \rangle$ for n-ary relations. Using a set of manually defined rules, contextual links between extracted tuples are inferred from the dependency parse in order to generate a nested representation of assertions that are complete and closer in meaning to the original statement. Similar to OLLIE, such contextual links are identified as clausal complements, conditionals and relation modifiers. Linked tuples are represented by arguments that refer to the corresponding extractions using identifiers, as shown in the following example (Bhutani et al., 2016):

#1: ⟨*body*; *appeared to have been thrown*; ∅⟩
#2: ⟨#1; *from*; *vehicle*⟩

Moreover, the example below denotes a temporal relationship between the extractions #1 and #2, based on the following sentence:

> "*After giving 5,000 people a second chance at life, doctors are celebrating the 25th anniversary of Britain's first heart transplant.*"

#1: ⟨*doctors*; *are celebrating*; *the 25th anniversary of Britain's first heart transplant*⟩
#2: ⟨*doctors*; *giving*; *second chance at life*⟩
#3: ⟨#1; *after*; #2⟩

MinIE (Gashteovski et al., 2017), another recent Open IE system, is built on top of ClausIE, a system that was found to often produce overspecified extractions. Such overly specific constituents that combine multiple, potentially semantically unrelated propositions in a single relational or argument phrase may easily hurt the performance of downstream NLP applications, such as QA or textual entailment. Instead, those approaches benefit from extractions that are as compact as possible. Therefore, MinIE aims to minimize both relational and argument phrases by identifying and removing parts that are considered overly specific. For this purpose, MinIE provides four different minimization modes which differ in their aggressiveness, thus allowing to control the trade-off between precision and recall. Moreover, it semantically annotates extractions with information about polarity, modality, attribution and quantities instead of directly representing it in the actual extractions, as the following example shows (Gashteovski et al., 2017):

> "*Pinocchio believes that the hero Superman was not actually born on beautiful Krypton.*"

#1: ⟨*Superman*; *was born actually on*; *beautiful Krypton*⟩
        Annotation: factuality, (- [not], certainty), attribution (Pinocchio, +, possibility [believes])

#2: ⟨*Superman*; *was born on*; *beautiful Krypton*⟩
Annotation: factuality, (- [not], certainty), attribution (Pinocchio, +, possibility [believes])
#3: ⟨*Superman*; *"is"*; *hero*⟩
Annotation: factuality, (+, certainty)

with + and − signifying positive and negative polarity, respectively.

Table 2.26 Properties of Open IE approaches capturing inter-proposition relationships

Approach	Features	Arity	Type of Relations
OLLIE	dependency parse	binary	mediated by verbs, nouns and adjectives
OPENIE- 4/5	• shallow syntax (part-of-speech tagging and noun phrase chunking) • constituency parse	n-ary	mediated by verbs, nouns and adjectives
CSD-IE	constituency parse	binary	verb- and noun-based
NESTIE	dependency parse	n-ary	verb- and noun-based
MinIE	dependency parse	binary	verb-based relations and noun-mediated relations from appositions and possessives

In that way, the output generated by MinIE is further reduced to its core constituents, producing maximally shortened, semantically enriched extractions.

Table 2.26 gives a summary of the properties of above-mentioned Open IE approaches that capture inter-proposition relationships by incorporating the semantic context between the extracted tuples. In that way, important contextual information that is needed to infer the true meaning of a statement is preserved, resulting in an output that increases the informativeness of the extracted relational tuples.

2.3.1.4 End-to-End Neural Approaches

With Stanovsky et al. (2018) and Cui et al. (2018), two end-to-end neural approaches for the task of Open IE are proposed. In Cui et al. (2018), Open IE is cast as a sequence-to-sequence generation problem, where the input sequence is the sentence and the output sequence consists of the corresponding relational tuples. Pairs of input and output sequences, serving as training instances for learning arguments and their corresponding relational phrases, are obtained from highly confident tuples that are bootstrapped from the OpenIE-4 system. Encoder and decoder are implemented using RNNs with an attention-based copying mechanism. In that way, the dependencies on other NLP tools (e.g., syntax parsers) are reduced, while bypassing hand-crafted patterns and alleviating error propagation. However, Cui et al. (2018)'s framework is restricted to the extraction of binary tuples, disregarding more complex structures such as n-ary and nested extractions, often resulting in long and complex argument phrases.

Similar to Cui et al. (2018), with RnnOIE, Stanovsky et al. (2018) present an encoder-decoder based Open IE approach. The authors reformulate Open IE as an SL task, using a custom BIO scheme (Ramshaw and Marcus, 1995) inspired by a recent state-of-the-art deep learning approach for SRL (He et al., 2017).

Table 2.27 Properties of end-to-end neural approaches to Open IE

Approach	Features	Arity	Type of Relations
Cui et al. (2018)	Seq2Seq (RNN)	binary	mediated by verbs, nouns and adjectives
RnnOIE	SL	n-ary	verb-based

Table 2.27 summarizes the properties of end-to-end neural approaches to the Open IE task.

2.3.2 Benchmarking Open Information Extraction Approaches

Though a multitude of systems for Open IE have been developed over the last decade, a clear formal specification of what constitutes a valid relational tuple is still missing. This lack of a well-defined, generally accepted task definition prevented the creation of an established, large-scale annotated corpus serving as a gold standard collection for an objective and reproducible cross-system comparison. As a consequence, to date, Open IE systems are predominantly evaluated by hand on small-scale corpora that consist of only a few hundred sentences, thereby ignoring one of the fundamental goals of Open IE: scalability to large amounts of text. Moreover, none of the datasets that were used for assessing the performance of different systems is widely agreed upon. As can be seen in Table 2.28, the corpora compiled by Del Corro and Gemulla (2013), Xu et al. (2013), Fader et al. (2011), Mesquita et al. (2013) and Banko et al. (2007) are occasionally re-used. However, new datasets are still collected, hindering a fair comparison of the proposed approaches. Besides, although Open IE methods are targeted at being domain independent and able to cope with heterogeneous datasets, the corpora used in the evaluation process are restricted to the news, Wikipedia and Web domains for the most part. Accordingly, no clear statement about the portability of the approaches to various genres of text is possible. In addition, most evaluation procedures described in the literature focus on precision-oriented metrics, while either completely ignoring recall or using some kind of proxy, such as yield, i.e. the total number of extractions labeled as correct, or coverage, i.e. the percentage of text from the input that is contained in at least one of the extractions.

Hence, the absence of a standard evaluation procedure makes it hard to replicate and compare the performance of different Open IE systems. Table 2.28 provides a detailed overview of both the datasets and the metrics that were used to evaluate the various approaches to the task of Open IE that were described above.

OIE2016 In order to address aforementioned difficulties, Stanovsky and Dagan (2016) recently made a first attempt in standardizing the Open IE evaluation by providing a large gold benchmark corpus. It is based on a set of consensual guiding principles that underlie most Open IE approaches proposed so far, as they have identified. Those principles cover the core aspects of the task of Open IE, allowing for a clearer formulation of the problem to be solved. The three key features to consider are the following:

Assertedness The assertedness principle states that each tuple extracted from the input should be asserted by the original sentence. Usually, instead of inferring predicate-argument structures from implied statements, e.g. the tuple ⟨*Sam*; *convinced*; *John*⟩ from ⟨*Sam*; *succeeded in convincing*; *John*⟩, Open IE systems tend to extract the full relational phrase (⟨*Sam*; *succeeded in convincing*; *John*⟩), incorporating matrix verbs ("*succeeded*") and other elements, such as negations or modals (e.g., ⟨*John*; *could not join*; *the band*⟩). Therefore, systems should not be penalized for not finding inferred relations.

Minimality In order to benefit downstream NLP tasks, it is favorable for Open IE systems to extract compact, self-contained tuples that do not combine several unrelated facts. Therefore, systems should aim to generate valid extractions with minimal spans for both relation and argument slots, while preserving the meaning of the input. For instance, the coordination in the sentence "*Bell distributes electronic and building products*" should ideally yield the following two

Table 2.28 Comparison of the approaches that were employed to evaluate the proposed Open IE frameworks. The first column indicates the respective Open IE system, while the second column denotes the size and domain of the datasets upon which the experiments were performed. The last column indicates the metrics that were used for assessing the systems' performance

Approach	# Sentences and Domain	Metrics
TEXT-RUNNER	400 Web	% correct extractions
WOE	• 300 news • 300 Wikipedia • 300 Web	precision-recall (PR) curve
REVERB	500 Web	PR curve
OLLIE	• 300 news (from WOE) • 300 Wikipedia (from WOE) • 300 biology	precision-yield curve
KRAKEN	500 Web (from REVERB)	• precision • completeness • # facts extracted per sentence

EXEMPLAR	• 500 Web (from TEXTRUNNER)
	• 500 news
	• 100 news (from Xu et al. (2013))
	• 222 news
		• binary:
		  − precision
		  − recall
		  − F_1-score
		• n-ary:
		  − precision over arguments
		  − recall over arguments

(a) Early Open IE approaches based on hand-crafted extraction patterns and self-supervised learning.

(Continued)

Table 2.28 (Continued)

Approach	# Sentences and Domain	Metrics
ClausIE	• 500 Web (from REVERB) • 200 Wikipedia • 200 news	• precision-yield curve • % correct extractions
Schmidek and Barbosa (2014)	• 500 Web (from TEXTRUNNER) • 500 news • 100 news (from Xu et al. (2013))	• precision • recall • F$_1$-score • time per sentence before and after sentence re-structuring
Stanford Open IE		extrinsic evaluation: TAC KBP Slot Filling Challenge (Surdeanu, 2013)

(b) Paraphrase-based Open IE approaches.

(Continued)

relational tuples: ⟨*Bell*; *distributes*; *electronic products*⟩ and ⟨*Bell*; *distributes*; *building products*⟩.

Completeness and Open Lexicon   The completeness and open lexicon principle aims to extract all relations that are asserted in the input text. This principle was one of the fundamental ideas that have been introduced in the work of Banko et al. (2007) together with the Open IE terminology. In their work, the Open IE task was defined as a domain-independent task which extracts all possible relations from heterogeneous corpora, instead of only extracting a set of pre-specified classes of relations. The majority of state-of-the-art Open IE systems realize this challenge by considering all possible verbs as potential relations. Accordingly, their scope is often limited to the extraction of verbal predicates, while ignoring relations mediated by more complex syntactic constructs, such as nouns or adjectives.

Realizing that above-mentioned requirements are subsumed by the task of QA-driven SRL (He et al., 2015), Stanovsky and Dagan (2016) converted the annotations

Table 2.28 (Continued)

Approach	# Sentences and Domain	Metrics
OPENIE-4	not reported	• precision • yield
CSD-IE	• 200 Wikipedia (from ClausIE) • 200 news (from ClausIE)	• % triples labeled accurate • % correct triples labeled minimal • coverage (% text contained in at least one triple) • average triple length
NESTIE	• 200 Wikipedia (from ClausIE) • 200 news (from ClausIE)	• correctness (0/1) • minimality (0/1) • informativeness (0-5)
MinIE	• 10k news (from Sandhaus (2008)) • 200 Wikipedia (from ClausIE) • 200 news (from ClausIE)	• # extractions • # non-redundant extractions • recall • factual precision • attribution precision • mean word count per triple (proxy for minimality)

(c) Open IE systems that capture inter-proposition relationships.

(Continued)

of a QA-SRL dataset into an Open IE corpus.[20] Using a rule-based system which asked questions about the source sentence with respect to each verbal predicate contained in the input (e.g. *"who __did__ something?"*, *"what __did__ she do?"*), the gold tuples were automatically created, expressing the Cartesian product of answers to the questions about each predicate. In that way, one relational tuple per predicate was generated, resulting in more than 10,000 extractions over 3,200 sentences from

---

[20] The OIE2016 benchmark dataset is available under https://github.com/gabrielStanovsky/oie-benchmark.

Table 2.28 (Continued)

Approach	# Sentences and Domain	Metrics
Cui et al. (2018)	3,200 news and Wikipedia (Stanovsky and Dagan, 2016)	• PR curve • area under the PR curve (AUC) • running time
RnnOIE	• 3,200 news and Wikipedia (Stanovsky and Dagan, 2016) • 3,300 wikinews and Wikipedia • 500 Web (from TEXTRUNNER) • 222 news (from EXEMPLAR) • 100 news (from Xu et al. (2013))	• PR curve • AUC • F_1-score

(d) End-to-end neural Open IE approaches.

Wikipedia and the Wall Street Journal. Supporting the calculation of the PR curve and the AUC, the OIE2016 benchmark allows for a quantitative analysis of the output of an Open IE system.

AW-OIE With AW-OIE ("All Words Open IE"), Stanovsky et al. (2018) present another Open IE corpus derived from Question-Answer Meaning Representation (QAMR) (Michael et al., 2018), an extension of the QA-SRL paradigm. Like QA-SRL, QAMR represents predicate-argument structures with a set of question-answer pairs for a given sentence, where each answer is a span from the input. However, while QA-SRL uses question templates that are centered on verbs, QAMR supports free-form questions over a wide range of predicate types. In that way, it allows to express richer, more complex relations, including for example relations mediated by nouns or adjectives. In total, the AW-OIE corpus consists of 17,165 tuples over 3,300 sentences from Wikipedia and Wikinews. It is intended to be used as an extension of the OIE2016 training set.

RelVis In addition, Schneider et al. (2017) compile RelVis, another benchmark framework for Open IE that allows for a large-scale comparative analysis of Open IE approaches. Besides Stanovsky and Dagan (2016)'s benchmark suite, it comprises the n-ary news dataset proposed in Mesquita et al. (2013), Banko et al. (2007)'s Web corpus and the Penn sentences from Xu et al. (2013). Similar to the toolkit proposed in Stanovsky and Dagan (2016), RelVis supports a quantitative evaluation of the performance of Open IE systems in terms of precision, recall and F_2-score. Furthermore, it facilitates a manual qualitative error analysis. For this purpose, six common error classes are distinguished to which inaccurate extractions can be assigned:

(1) *wrong boundaries*, where the relational or argument phrase is either too long or too small;
(2) *redundant extraction*, where the proposition asserted in an extraction is already expressed in another extraction;
(3) *uninformative extraction*, where critical information is omitted;
(4) *missing extraction*, i.e. a false negative, where either a relation is not detected by the system or the argument-finding heuristics choose the wrong arguments or no argument at all;
(5) *wrong extraction*, where no meaningful interpretation of the extracted relational tuple is possible; and
(6) *out of scope extraction*, where a system yields a correct extraction that was not recognized by the authors of the gold dataset.

WiRe57 With WiRe57, Lechelle et al. (2019) propose a high-quality manually curated Open IE dataset.[21] It was created by two Open IE experts who annotated the set of relations contained in 57 sentences taken from three Wikipedia and two newswire articles, resulting in a ground truth reference of 343 relational tuples. In accordance with the key features of the task of Open IE identified in Stanovsky and Dagan (2016), they adhere to the following guiding principles in the annotation process:

- The extracted information should be *informative*.
- The extracted tuples should be *minimal*, i.e. they should convey the smallest stand-alone piece of information, though that piece must be completely expressed.

---

[21] The WiRe57 Open IE benchmark can be downloaded from https://github.com/rali-udem/WiRe57.

- The annotation shall be *exhaustive*, i.e. it should capture as much of the information expressed in the input sentence as possible.
- Information that is not explicitly expressed in the input, but can only be *inferred* from it, should not be annotated.

The WiRe57 benchmark assigns a token-level precision and recall score to all gold-prediction pairs of a sentence, ignoring the predictions' confidence values that allow for the computation of the PR curve in the OIE2016 benchmark. Moreover, with only 343 reference tuples, this dataset is too small to suffice as a test dataset for a reliable and comprehensive comparative analysis of the performance of Open IE approaches.

CaRB Arguing that previously proposed Open IE benchmarks are either too small or too noisy to allow for a thorough and meaningful comparison of different Open IE approaches, Bhardwaj et al. (2019) compile CaRB, a crowdsourced benchmark for Open IE ("**C**rowdsourced **a**utomatic Open **R**elation extraction **B**enchmark").[22] Using Amazon Mechanical Turk, workers are asked to annotate the 1,282 sentences from the development and test sets of the OIE2016 benchmark, following a set of annotation guidelines similar to the principles in Lechelle et al. (2019):

- *Completeness*: The workers must attempt to extract *all* assertions from the input sentence.
- *Assertedness*: Each tuple must be implied by the original sentence. Though as opposed to Stanovsky and Dagan (2016) and Lechelle et al. (2019), the relations do not need to be explicitly expressed in the input sentence, but can also be inferred from it.
- *Informativeness*: Workers must include the maximum amount of relevant information in an argument.
- *Atomicity*: In accordance with the minimality principle of Stanovsky and Dagan (2016) and Lechelle et al. (2019), each tuple must be an indivisible unit.

In that way, a large-scale, high-quality dataset of 5,263 ground truth tuples is created. Similarly to the OIE2016 benchmark, it allows for a comparative analysis of the output of Open IE systems based on their PR curves and AUC scores. However, an improved approach for matching predicted extractions to reference tuples is used (for details see Section 14.3.3.1).

---

[22] The CaRB benchmark for Open IE is available for download under https://github.com/dair-iitd/CaRB.

OPIEC   Comprising 340 million relational tuples, OPIEC is by far the largest Open
IE corpus publicly available to date (Gashteovski et al., 2019).[23] It was automatically
extracted from the full text of the English Wikipedia using the state-of-the-art Open
IE system MinIE. However, a detailed analysis of the resulting dataset revealed that
it is very noisy, since a large part of the extractions are either underspecified in that
additional information from the input is required to obtain a coherent piece of information, or overly specific, e.g. because of arguments consisting of complex phrases.
Thus, to remove such under-specific and complex tuples, the OPIEC-Clean subcor-
pus was created, in which only those extractions that express relations between
entities or concepts were kept. In that way, the original dataset was reduced to 104
million extractions, with 66% of the tuples having at least one argument that is a
named entity. However, restraining arguments to be named entities limits the task of
IE to capturing only the most salient relations expressed in the text (Lechelle et al.,
2019).

Table 2.29 Summary of the properties of the benchmarks for the evaluation of Open IE
approaches. AW-OIE and OPIEC do not specify an evaluation process, but only provide
corpora of input sentences and corresponding extractions

Dataset	Domain	#sentences	#tuples	Evaluation metrics
OIE2016	news, Wikipedia	3,200	10,359	AUC, PR curve
AW-OIE	Wikinews, Wikipedia	3,300	17,165	–
RelVis	news, Wikipedia, Web, mixed	4,022	11,093	Precision, Recall, F_2, qualitative manual analysis
WiRe57	news, Wikipedia	57	343	Precision, Recall, F_1
CaRB	news, Wikipedia	1,282	5,263	AUC, PR curve
OPIEC	Wikipedia	full English Wikipedia	340 million	–
OPIEC-Clean	Wikipedia	full English Wikipedia	104 million	–

---

[23] The OPIEC corpus is available under https://www.uni-mannheim.de/dws/research/
resources/opiec/.

To put it in a nutshell, until recently, there was not a single gold standard dataset over which Open IE systems were evaluated. Instead, their performance has traditionally been assessed on small, proprietary datasets that were manually curated. This lack of a universally accepted ground truth dataset impeded the comparison of results among different Open IE systems in a large-scale, objective and reproducible fashion.

It was only recently that the first moves towards a standardized mechanism for an automatic evaluation of Open IE systems were made by compiling benchmark frameworks that operate on a larger scale. Since its release in 2016, the OIE2016 benchmark has become the de facto standard for the evaluation of Open IE approaches. Other frameworks, including RelVis, WiRe57, CaRB and OPIEC, have been proposed since then, trying to overcome some of the limitations identified in the OIE2016 benchmark (see Section 14.3.3.1). Table 2.29 provides an overview of the characteristics of the Open IE benchmarks presented above.

2.4 Meaning Representations

According to Del Corro and Gemulla (2013), Open IE is "perhaps the simplest form of a semantic analysis." As described in the previous section, Open IE aims to obtain a shallow semantic representation of texts in the form of predicates and their arguments. Such predicate-argument structures are regarded as fundamental components of a semantic representation of a sentence, which is often also called a MR. Thus, the task of Open IE can be seen as a first step towards a richer semantic analysis (Del Corro and Gemulla, 2013).

Schemes for the semantic representation of text aim to reflect the meaning of sentences in a transparent way, by assigning similar structures to different constructions that share the same basic meaning and assigning different structures to constructions that have different meanings, despite their surface similarity. The fundamental component conveyed by semantic schemes is the predicate-argument structure that identifies events, their participants and the relations between them by specifying *who* did *what* to *whom*, *where* and *why* (Màrquez et al., 2008). However, they largely diverge in their organizing principles, granularity, event types and types of predicates that they cover, as well as their cross-linguistic applicability and their relation with syntax. In the following, we will briefly survey the most widely used schemes for the semantic representation of text.

SRL   The task of SRL is closely related to Open IE. The main difference is that the former aims not only to identify the boundaries of the arguments of a predicate (*argument identification*), but also to label them with their semantic roles (*argument classification*). In that way, SRL schemes are able to indicate exactly what types of semantic relations hold among a predicate and its associated participants and properties. Typical roles used in SRL are labels such as Agent, Patient and Location for entities participating in an event, as well as Temporal and Manner for the characterization of other aspects of the event or participant relations, as illustrated below (Màrquez et al., 2008; Kearns, 2011):

- "*[The girl on the swing]*$_{\text{AGENT}}$ *[whispers]*$_{\text{PREDICATE}}$ *[to the boy beside her]*$_{\text{RECIPIENT}}$."
- "*[The jug]*$_{\text{THEME}}$ *[remained]*$_{\text{PREDICATE}}$ *[on the table]*$_{\text{LOCATION}}$."
- "*[Tom]*$_{\text{AGENT}}$ *[served]*$_{\text{PREDICATE}}$ *[Sally]*$_{\text{RECIPIENT}}$ *[spaghetti]*$_{\text{THEME}}$ *[with a silver spoon]*$_{\text{INSTRUMENT}}$."

There is no consensus on a definitive list of semantic roles (also called *"thematic roles"*). Instead, the roles that the different resources encompass vary greatly. The leading SRL schemes are FrameNet (Fillmore et al., 2004), VerbNet (Kipper et al., 2000) and PropBank (Palmer et al., 2005).

AMR   Abstract Meaning Representation (AMR) (Banarescu et al., 2013) is a semantic scheme that covers predicate-argument relations for a wide variety of predicate types, including verbal, nominal and adjectival predicates. Analogous to SRL schemes, it also identifies the semantic roles of the arguments. For this purpose, it makes use of PropBank framesets. It is represented as a rooted, directed graph whose leaves are labeled with concepts, which are either English words, PropBank framesets or special keywords, such as entity types, quantities or logical conjunctions. The graph nodes are connected through relations. In total, AMR distinguishes about 100 different types of relations, including frame arguments that follow Prop-Bank conventions (e.g., *"arg0"*, *"arg1"*, *"arg2"*), general semantic relations (e.g., *"age"*, *"cause"*, *"direction"*), relations for quantities, relations for date-entities and relations for lists.

UCCA   UCCA (Abend and Rappoport, 2013) is a semantic scheme that is based on typological (Dixon, 2010a; Dixon, 2010b; Dixon, 2012) and cognitive (Langacker, 2008) theories (Sulem et al., 2018b). It uses directed acyclic graphs to represent semantic structures and is built as a multi-layered structure, which allows for open-ended extension. Its foundational layer provides a coarse-grained representation where a text is regarded as a collection of so-called Scenes. A Scene describes a movement, an action or a state which persists in time. Every Scene contains one main relation, which can be either a Process (describing a temporally evolving event) or a State (describing a temporally persistent state). Moreover, a Scene contains one or more Participants, subsuming both concrete and more abstract entities. In addition, UCCA annotates inter-Scene relations, distinguishing between three major types of linkage: a Scene can be a Participant in another Scene; a Scene may provide additional information about an established entity (*"Elaborator Scene"*); or a Scene may happen in parallel to other Scenes, expressed through for example temporal, causal or conditional relationships (*"Parallel Scene"*). As opposed to the English-centric AMR, UCCA is cross-linguistically applicable.

Decomp   Universal Decompositional Semantics (Decomp) (White et al., 2016) aims to augment Universal Dependencies (de Marneffe et al., 2014) data sets with semantic annotations that scale across different types of semantic information and different languages. It is a multi-layered scheme that supports the annotation of semantic role information, word senses and aspectual classes (e.g., realis/irrealis). For this purpose, Decomp incorporates semantic decomposition directly into its protocols, by mapping from decompositional theories (e.g., Dowty (1991)) into straightforward questions on binary properties that are easily answered.

DRT   Discourse Representation Theory (DRT) is a further well-known theory of MR (Kamp, 1981). It is based on the idea that the interpretation of the current sentence is dependent on the interpretation of the sentences that precede it. Hence, in DRT, semantic interpretation is regarded as a dynamic process, causing a shift from static to dynamic semantics (Lascarides and Asher, 2007). As new sentences are added to an existing piece of interpreted discourse, the representation of the discourse is continuously updated. In this context, DRT focuses in particular on the effect of logical structure on anaphora of various kinds, such as pronouns (Kamp and Reyle, 1993), tense (Kamp and Rohrer, 1983) and presupposition (Van der Sandt, 1992). The basic meaning-carrying units in DRT are DRSs, which consist of two components: first, discourse referents (e.g., x, y) that represent entities in the discourse and second, discourse conditions, representing information about discourse referents (Liu et al., 2019). The latter can be atoms, links or complex conditions. An atom is a predicate name applied to a number of discourse referents (e.g., $man(x)$, $beer(z)$, $order(y, z)$), while a link is an expression $y = x$, where x is a discourse referent and y is a proper name or discourse referent. Complex conditions, finally, allow for the use of quantifiers and logical operators, including conditionals and negations (e.g., $man(x) \rightarrow walk(x)$) (van Eijck, 1990). DRT has been further developed and improved over the years. Segmented Discourse Representation Theory (SDRT) (Lascarides and Asher, 2007), for example, extends the language of DRSs by introducing rhetorical relations into the semantic representation of the discourse, thus augmenting its compositional semantics.

Part II
Discourse-Aware Sentence Splitting

Introduction

Sentences that present a complex linguistic structure can be hard to comprehend by human readers, as well as difficult to analyze by NLP applications (Mitkov and Saggion, 2018). Identifying grammatical complexities in a sentence and transforming them into simpler structures is the goal of syntactic TS. One of the major types of operations that are used to perform this rewriting step is sentence splitting: it divides a sentence into several shorter components, with each of them presenting a simpler and more regular structure that is easier to process by both humans (Siddharthan and Mandya, 2014; Saggion et al., 2015; Ferrés et al., 2016) and machines (Štajner and Popovic, 2016; Štajner and Popović, 2018; Saha and Mausam, 2018).

Based on the research gaps identified in Section 1.3, we develop a syntactic TS approach that focuses on the task of sentence splitting and thus the type of rewriting operation that has been ignored to a large extent in the TS research so far. The framework that we propose is intended to serve as a preprocessing step to generate an intermediate representation of the input that facilitates and improves the performance of downstream Open IE tasks. The goal of our approach is to break down a complex source sentence into a set of *minimal propositions*, i.e. a sequence of sound, self-contained utterances, with each of them presenting a single event that cannot be further decomposed into meaningful propositions (see Section 4.1). In that way, we aim to overcome the conservatism exhibited by state-of-the-art syntactic TS approaches, i.e. their tendency to retain the input rather than transforming it into a simplified output.

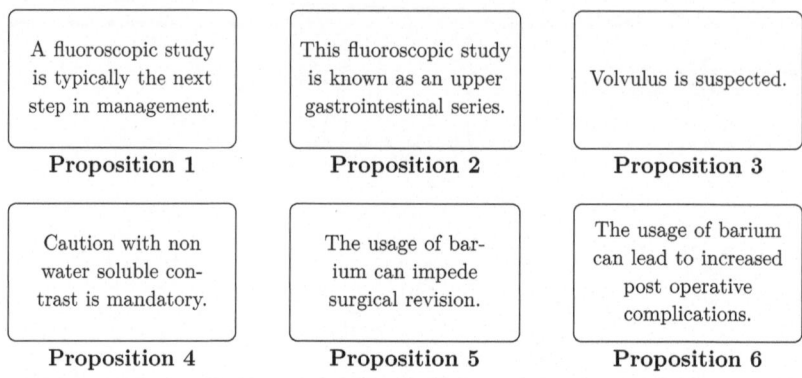

A fluoroscopic study is typically the next step in management.	This fluoroscopic study is known as an upper gastrointestinal series.	Volvulus is suspected.
Proposition 1	**Proposition 2**	**Proposition 3**
Caution with non water soluble contrast is mandatory.	The usage of barium can impede surgical revision.	The usage of barium can lead to increased post operative complications.
Proposition 4	**Proposition 5**	**Proposition 6**

Figure 3.1 Minimal propositions resulting from the sentence splitting subtask on our running example from Table 1.1. The input is decomposed into a loose sequence of minimal semantic units that lacks coherence

However, any sound and coherent text is not simply a loose arrangement of self-contained units, but rather a logical structure of utterances that are semantically connected (Siddharthan, 2014). Consequently, when carrying out syntactic TS operations without considering discourse implications, the rewriting may easily result in a disconnected sequence of simplified sentences, making the text harder to interpret, as can be witnessed in the output depicted in Figure 3.1. The vast majority of existing structural TS approaches though do not take into account discourse-level aspects (Alva-Manchego et al., 2020). Therefore, they are prone to producing a set of incoherent utterances where important contextual information is lost. Consequently, in particular in case of complex assertions, the generated output commonly lacks the expressiveness that is needed for a proper interpretation in downstream Open IE applications. Thus, in order to preserve the coherence structure and, hence, the interpretability of the output, we propose a discourse-aware TS approach based on the framework of RST. By distinguishing between core and context information, our approach establishes a *contextual hierarchy* between the split components. In addition, it identifies and classifies the *semantic relationship* that holds between them. In that way, a semantic hierarchy is set up between the decomposed spans. See Figure 3.2 for an example of the resulting context-preserving output, consisting of a set of hierarchically ordered and semantically interconnected propositions.

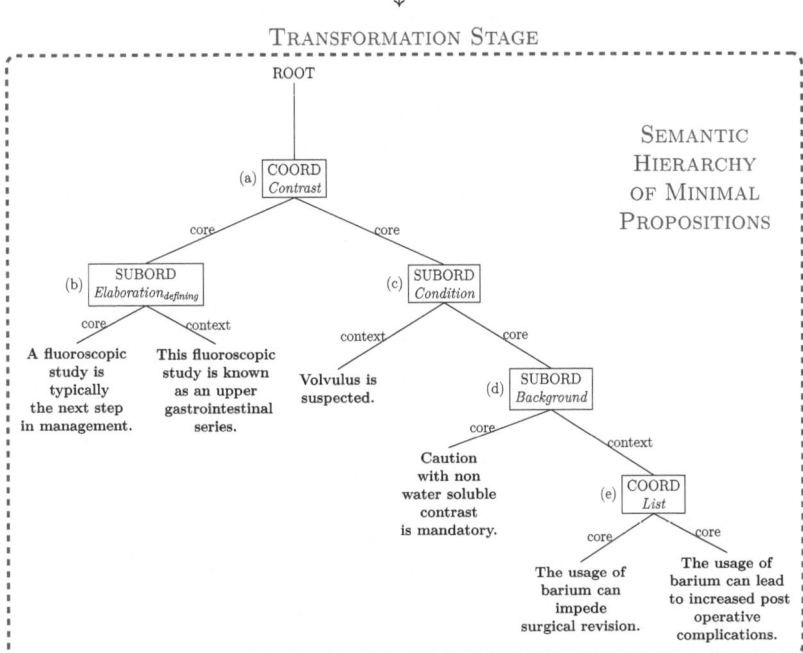

Input: Complex Sentence

A fluoroscopic study which is known as an upper gastrointestinal series is typically the next step in management, although if volvulus is suspected, caution with non water soluble contrast is mandatory as the usage of barium can impede surgical revision and lead to increased post operative complications.

Transformation Stage

ROOT

Semantic Hierarchy of Minimal Propositions

(a) COORD *Contrast*

core   core

(b) SUBORD *Elaboration_{defining}*

core   context

A fluoroscopic study is typically the next step in management.

This fluoroscopic study is known as an upper gastrointestinal series.

(c) SUBORD *Condition*

context   core

Volvulus is suspected.

(d) SUBORD *Background*

core   context

Caution with non water soluble contrast is mandatory.

(e) COORD *List*

core   core

The usage of barium can impede surgical revision.

The usage of barium can lead to increased post operative complications.

Figure 3.2 Semantic hierarchy of minimal propositions representing the context-preserving output that is generated by our discourse-aware TS approach on the source sentence from Table 1.1. A complex input sentence is transformed into a semantic hierarchy of simplified sentences in the form of minimal, self-contained propositions that are linked to each other via rhetorical relations. The output presents a regular, fine-grained structure that preserves the semantic context of the input, allowing for a proper interpretation of complex assertions

Subtask 1: Splitting into Minimal Propositions

4

The major objective of our TS approach is to transform sentences that present complex structures into a set of *easy-to-access sentences* which will be easier for downstream Open IE applications to deal with due to their simpler and more regular syntax (Beigman Klebanov et al., 2004). During the transformation process, the following three properties need to be ensured for the resulting simplified sentences: (i) *syntactic correctness*, (ii) *semantic correctness* and (iii) *minimality* (see Figure 4.1). While the first two features are rather straightforward to characterize, the latter requires a more precise definition that will be detailed in the following section. In Section 4.2, we then present our approach for splitting complex source sentences into a set of structurally simplified sentences.

Supplementary Information The online version contains supplementary material available at https://doi.org/10.1007/978-3-658-38697-9_4.

C. Niklaus, *From Complex Sentences to a Formal Semantic Representation using Syntactic Text Simplification and Open Information Extraction*,
https://doi.org/10.1007/978-3-658-38697-9_4

SUBTASK 1: SPLIT INTO MINIMAL PROPOSITIONS

Input: A complex sentence C.

Problem: Produce a sequence of simple sentences $T_1, \ldots, T_n$, $n >= 2$, such that

1. each simple sentence T_i is grammatically sound. (**syntactic correctness**)
2. the output sentences $T_1, \ldots, T_n$ convey all and only the information in C. (**semantic correctness**)
3. each simple sentence T_i presents a minimal semantic unit, i.e. it cannot be further decomposed into meaningful propositions. (**minimality**)

Figure 4.1 Problem description of subtask 1, whose goal is to split syntactically complex sentences into a set of minimal propositions

4.1 Property of Minimality

Our approach addresses sentence-level factual TS, i.e. it is restricted to declarative sentences, which can be used to make statements that are either true or false (Kearns, 2011). The framework we propose takes as input sentences with a complex syntax and splits them step-by-step into a set of structurally simplified sentences. Intuitively, the generated output is *"easy to read and understand, and arguably easily processed by computers"* (Bach et al., 2011). In the following, we present a more precise specification of the minimality property we aim for in the resulting simplified sentences, both on a syntactic and a semantic level.

4.1.1 Minimality on the Syntactic Level

In syntax, four types of sentence structures are distinguished (Quirk et al., 1985):

(1) **simple sentence**: A simple sentence contains only *one independent clause*. An independent clause is a group of words that has both a subject and a verb. It expresses a complete thought, i.e. some coherent piece of information (Del Corro and Gemulla, 2013), as the following example shows:

"I admire her reasoning."

(2) **compound sentence**: A compound sentence is composed of at least *two inde-pendent clauses* which are joined by a coordinating conjunction or a semicolon, e.g.:

"I admire her reasoning, but I reject her conclusions."

(3) **complex sentence**: A complex sentence contains *a subordinate clause and an independent clause*. Similarly to independent clauses, a subordinate clause is a group of words that has both a subject and a verb. However, it does not express a complete thought. For instance, in the example below, the adver-bial clause *"Although I admire her reasoning"* represents a subordinate clause which depends on the superordinate independent clause *"I reject her conclu-sions"*, modifying the verb *"reject"*:

"Although I admire her reasoning, I reject her conclusions."

(4) **compound-complex sentence**: As the name suggests, a compound-complex sentence is a *combination of compound and complex sentences*, containing at least two independent clauses and at least one subordinate clause, as illustrated below:

"I admire her reasoning, but I reject her conclusions since one of her assumptions is wrong."

Note that for the sake of simplicity, we do not strictly adhere to the syntactic notion of a complex sentence in the description of our TS approach. Instead, we subsume any sentence that does not present a minimal semantic unit (as detailed below) under the term of a complex sentence.

Based on the syntactic theory described above, the goal of our approach is to transform a given source sentence into a set of simple sentences, where each of them comprises exactly one independent clause that consists of one subject (S), one verb (V) and optionally an indirect object (O_i), a direct object (O) and one or more adverbials (A). For instance, consider the following example sentence (Del Corro and Gemulla, 2013):

"Albert Einstein died in Princeton in 1955, after he had refused further medical attention."

This sentence is composed of the following two independent clauses, representing so-called simple sentences:

- *"[Albert Einstein]$_S$ [died]$_V$ [in Princeton]$_A$ [in 1955]$_A$."*
- *"[He]$_S$ [had refused]$_V$ [further medical attention]$_O$."*

A closer look at the simple sentences of the above example reveals that there is still potential for further simplification. While the second sentence is limited to essential components, i.e. none of them can be omitted without making the sentence ill-formed, the former includes optional constituents that render it overly complex. In fact, the two adverbials *"in Princeton"* and *"in 1955"* specify additional contextual information that can be left out without producing a malformed output. Rather, the remaining clause *"Albert Einstein died"* still carries semantically meaningful information. Hence, in order to transform simple sentences into atomic propositions that cannot be further decomposed into meaningful units, we need to further reduce them to their corresponding clause type (Del Corro and Gemulla, 2013).

According to Quirk et al. (1985), clauses can be classified into seven different clause types based on the grammatical function of their constituents, as illustrated in Table 4.1 (Del Corro and Gemulla, 2013).

Table 4.1 The seven types of clauses. S: Subject, V: Verb, C: Complement, O: Object, A: Adverbial

	CLAUSE TYPE	EXAMPLE
T_1	SV	*[Albert Einstein]$_S$ [died]$_V$.*
T_2	SVA	*[Albert Einstein]$_S$ [remained]$_V$ [in Princeton]$_A$.*
T_3	SVC	*[Albert Einstein]$_S$ [is]$_V$ [smart]$_C$.*
T_4	SVO	*[Albert Einstein]$_S$ [has won]$_V$ [the Nobel Prize]$_O$.*
T_5	SVOO	*[The Royal Swedish Academy of Sciences]$_S$ [gave]$_V$ [Albert Einstein]$_O$ [the Nobel Prize]$_O$.*
T_6	SVOA	*[The doorman]$_S$ [showed]$_V$ [Albert Einstein]$_O$ [to his office]$_A$.*
T_7	SVOC	*[Albert Einstein]$_S$ [declared]$_V$ [the meeting]$_O$ [open]$_C$.*

The clause type[1] conveys the *minimal unit of coherent information* in the clause. Accordingly, if a constituent of a clause that is also part of its type is removed, the resulting clause does not carry semantically meaningful information any more (or the sense of the verb changes) (Del Corro and Gemulla, 2013). Hence, while constituents that belong to the clause type are essential components of the corresponding simple sentence, all other constituents are optional and can be discarded without leading to an incoherent or semantically meaningless output.

So far, a minimal proposition can be defined as a simple sentence that is reduced to its clause type by omitting all optional constituents, i.e. all elements that do not appear in the type of the underlying clause. However, transforming complex source sentences into simple sentences and trimming them to their clause types may still result in over-specified propositions due to overly complex subclausal units, i.e. phrases, as the following example illustrates (Del Corro and Gemulla, 2013):

> *"Bell, a telecommunication company, which is based in Los Angeles, makes and distributes electrical goods, computers and building products."*

The sentence above joins two independent clauses that can be split into the following simple sentences:

- *"[Bell, a telecommunication company,]$_S$ [makes and distributes]$_V$ [electrical goods, computers and building products]$_O$."* of type SVO
- *"[Bell]$_S$ [is based]$_V$ [in Los Angeles]$_A$."*[2] of type SVA

In the example above, the phrasal elements in all three positions of the first simple sentence are unnecessarily complex. While the subject contains an appositive phrase that further specifies the noun to which it refers (*"Bell"*), both the verb and the object include coordinated conjunctions that can be decomposed into separate elements, resulting in a set of clauses that present a much simpler syntax. Consequently, we further simplify such utterances by extracting phrasal elements from the input and transforming them into stand-alone sentences. For instance, we generate synthetic clauses for appositions by introducing an artificial verb such as *"is"* and linking them with the phrase to which they refer. Moreover, coordinate verb and noun phrases

---

[1] The type of a clause is uniquely identified by the verb type along with the presence of a direct object, an indirect object or a complement (Del Corro and Gemulla, 2013).

[2] We replaced the relative pronoun *"which"* by the antecedent *"Bell"* on which the relative clause depends.

are simplified by replacing the coordinated conjunctions by each of the conjoints.[3] In that way, we avoid over-specified subclausal elements, resulting in the following simplified output for our example sentence:

- *"[Bell]$_S$ [is]$_V$ [a telecommunication company]$_A$."* of type SVA
- *"[Bell]$_S$ [makes]$_V$ [electrical goods]$_O$."* of type SVO
- *"[Bell]$_S$ [makes]$_V$ [computers]$_O$."* of type SVO
- *"[Bell]$_S$ [makes]$_V$ [building products]$_O$."* of type SVO
- *"[Bell]$_S$ [distributes]$_V$ [electrical goods]$_O$."* of type SVO
- *"[Bell]$_S$ [distributes]$_V$ [computers]$_O$."* of type SVO
- *"[Bell]$_S$ [distributes]$_V$ [building products]$_O$."* of type SVO
- *"[Bell]$_S$ [is based]$_V$ [in Los Angeles]$_A$."* of type SVA

In a similar way, our approach handles a further set of phrasal units which we identified to commonly contribute to overly complex structures, mixing multiple semantic units. These include participial phrases, prepositional phrases, adjectival/adverbial phrases and leading noun phrases (see Section 6.2). In that way, we transform source sentences that present a complex syntax into output sentences with a simple and regular structure that is easy to process by downstream Open IE applications. To sum up, by the notion of a minimal proposition we understand a simple sentence that has been broken down to its essential constituents, i.e. the elements that are part of its clause type, while also extracting a specified set of phrasal units.[4]

4.1.2 Minimality on the Semantic Level

Syntactic structures, as expatiated above, reflect first and foremost the formal constructions used for expressing meanings. Semantics, on the contrary, abstract away

---

[3] In noun phrases, the replacement of a coordinated conjunction by one of its conjoints may lead to incorrect simplifications in case of a combinatory (instead of a segregatory) coordination, e.g. *"Anna and Bob married each other."* (Del Corro and Gemulla, 2013). Currently, our approach does not handle this issue.

[4] Note that in theory, further simplification is possible, for example by decomposing attributive adjectives such as in *"The **black** dog is barking."* or by extracting adverbial modifiers such as in *"He will slice the Salami **carefully.**"* However, we consider the atomic propositions as previously defined as sufficiently simplified, since they already present a simple and regular structure that is easy to analyze by machines. Further simplifications would only increase the risk of introducing errors in the transformation process without leading to an output that is substantially more useful for downstream Open IE tasks. Therefore, we refrain from simplifying any additional grammatical structures.

from specific syntactic constructions (Abend and Rappoport, 2013). Accordingly, a semantic analysis aims to assign similar meaning representations to sentences that differ in their structure, yet share the same basic meaning (Abend and Rappoport, 2017).

From a semantic point of view, a text can be seen as a collection of *events* (also called "frames" or "scenes") (Abend and Rappoport, 2013) that describe some activity, state or property. Thus, the definition of an event is similar to the semantic aspect of a clause in Basic Linguistic Theory (Dixon, 2010a). The goal of a semantic analysis is to identify and characterize such events in a given sentence, by determining for example *who* did *what* to *whom*, *where*, *when* and *how* (Màrquez et al., 2008). For instance, consider the following example (Abend and Rappoport, 2013):

> *"Golf became a passion for his oldest daughter: she took daily lessons and became very good, reaching the Connecticut Golf Championship."*

This sentence contains four events, evoked by *"became a passion"*, *"took daily lessons"*, *"became very good"* and *"reaching"*. Consequently, it can be decomposed into the following individual frames:

- *"[Golf]*$_{\text{ARG}}$ *[became a passion]*$_{\text{PRED}}$ *[for his oldest daughter]*$_{\text{ARG}}$.*"
- *"[She]*$_{\text{ARG}}$ *[took daily lessons]*$_{\text{PRED}}$.*"
- *"[She]*$_{\text{ARG}}$ *[became very good]*$_{\text{PRED}}$.*"
- *"[She]*$_{\text{ARG}}$ *[was reaching]*$_{\text{PRED}}$ *[the Connecticut Golf Championship]*$_{\text{ARG}}$.*"

Thus, from a semantic perspective, the goal of our TS approach is to split sentences into separate frames, where each of them presents a single event.

An event typically consists of a predicate, a set of arguments and optionally secondary relations. While the predicate (typically a verb, but nominal or adjectival predicates are also possible) is the main determinant of what the event is about, arguments describe its participants (e.g., *who* and *where*). They represent core elements of a frame, i.e. essential components that make it unique and different from other frames. Secondary relations, in contrast, represent non-core elements that introduce additional relations, describing further event properties (e.g., *when* and *how*) (Abend and Rappoport, 2017, Màrquez et al., 2008), as the sentence below illustrates (Dixon, 2010a):

> *"[John]*$_{\text{CORE ARG}}$ *[hit]*$_{\text{PRED}}$ *[the vase]*$_{\text{CORE ARG}}$ *[with a stick]*$_{\text{NON-CORE ARG}}$.*"[5]

---

[5] representing an SVOA pattern of clause type SVO

The verb *"hit"* represents the predicate of the event, while the subject *"John"* and the object *"the vase"* represent its core arguments. The instrumental adverbial phrase *"with a stick"*, by contrast, acts as a peripheral argument. In order to avoid over-specified propositions that are difficult to handle for downstream Open IE applications, such non-core elements of a frame are extracted and transformed into stand-alone sentences in our proposed TS approach, resulting in the following simplified sentences for the above example:[6]

- *"[John]*CORE ARG *[hit]*PRED *[the vase]*CORE ARG*."*[7]
- *"[This]*CORE ARG *[was]*PRED *[with a stick]*CORE ARG*."*[8]

Finally, to prevent overly complex argument and predicate spans, coordinate verb and noun phrases are decomposed into separate frames as well.

To sum up, our TS approach aims to split sentences that present a complex structure into its basic semantic building blocks in the form of events. Complex sentences that contain several events are decomposed into separate frames, while non-core elements of a frame, as well as coordinated conjunctions, are extracted and transformed into stand-alone sentences. Hence, from a semantic point of view, a minimal proposition can be seen as an utterance expressing a single event consisting of a predicate and its core arguments.

4.2   Splitting Procedure

With the objective of transforming complex sentences into a set of minimal propositions - in accordance with the specifications from Section 4.1 -, source sentences that present a complex linguistic form are converted into simpler and more regular structures by disembedding clausal and phrasal components that contain only information.

For this purpose, our framework takes as input a sentence and performs a recursive transformation stage that is based upon a small set of 35 manually defined grammar rules.[9] In the development of these patterns, we followed a principled and

---

[6] Note that due to the reasons mentioned previously in Section 4.1.1, attributive adjectives as well as adverbial modifiers are not extracted from a given frame, though representing non-core elements.

[7] representing an SVO pattern of clause type SVO

[8] representing an SVA pattern of clause type SVA

[9] For reproducibility purposes, the complete set of transformation patterns is detailed in Section 6.

systematic procedure, with the goal of eliciting a universal set of transformation rules for converting complex source sentences into a set of minimal propositions. The patterns were heuristically determined in a rule-engineering process that was carried out on the basis of an in-depth study of the literature on syntactic sentence simplification (Siddharthan, 2006; Siddharthan, 2014; Siddharthan, 2002; Siddharthan and Mandya, 2014; Evans, 2011; Evans and Orăsan, 2019; Heilman and Smith, 2010; Vickrey and Koller, 2008; Shardlow, 2014; Mitkov and Saggion, 2018; Mallinson and Lapata, 2019; Brouwers et al., 2014; Suter et al., 2016; Ferrés et al., 2016; Chandrasekar et al., 1996). Next, we performed a thorough linguistic analysis of the syntactic phenomena that need to be tackled in the sentence splitting task. Details on the underlying linguistic principles, supporting the systemacity and universality of the developed transformation patterns, can be found in Section 6.

Table 4.2 Linguistic constructs addressed by our discourse-aware TS approach, including both the hierarchical level (see subtask 2a, Section 5.1) and the number of simplification rules that were specified for each syntactic construct

	CLAUSAL/PHRASAL TYPE	HIERARCHY	# RULES
Clausal disembedding			
1	Coordinate clauses	coordinate	1
2	Adverbial clauses	subordinate	6
3a	Relative clauses (non-restrictive)	subordinate	5
3b	Relative clauses (restrictive)	subordinate	4
4	Reported speech	subordinate	4
Phrasal disembedding			
5	Coordinate verb phrases	coordinate	1
6	Coordinate noun phrases	coordinate	2
7	Participial phrases	subordinate	4
8a	Appositions (non-restrictive)	subordinate	1
8b	Appositions (restrictive)	subordinate	1
9	Prepositional phrases	subordinate	3
10	Adjectival and adverbial phrases	subordinate	2
11	Lead noun phrases	subordinate	1
	Total		35

Table 4.2 provides an overview of the linguistic constructs that are addressed by our approach, including the number of patterns that were specified for the respective syntactic phenomenon.

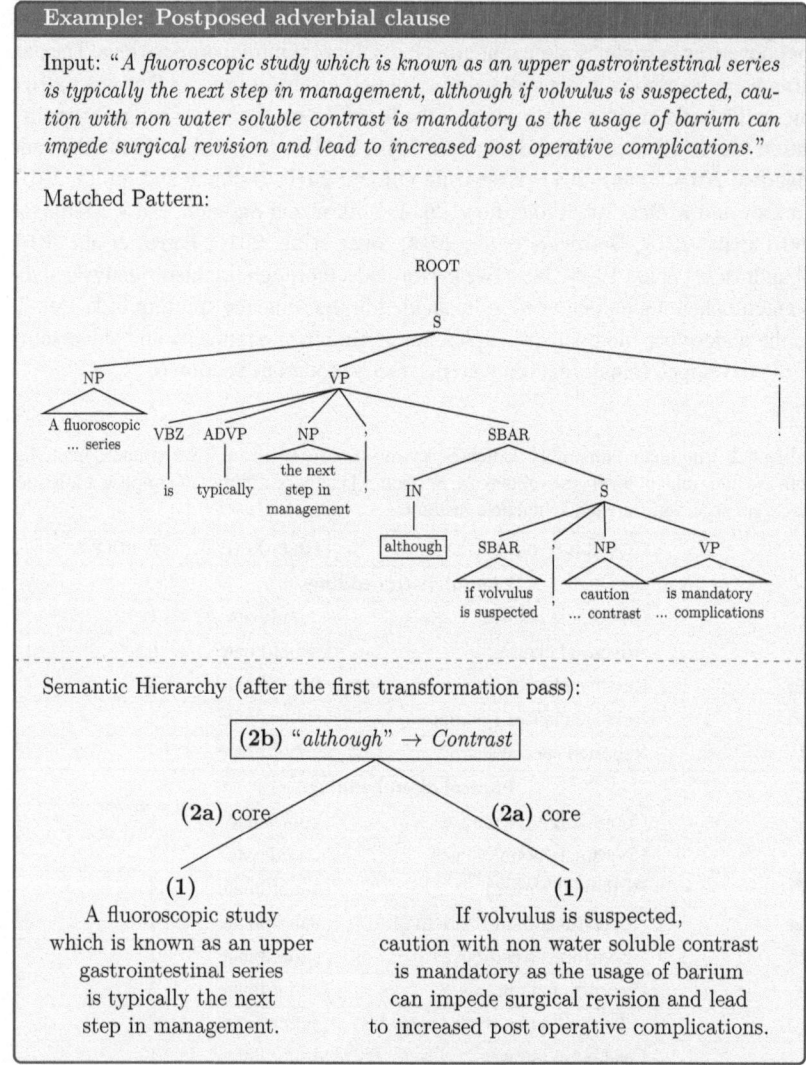

Example: Postposed adverbial clause

Input: *"A fluoroscopic study which is known as an upper gastrointestinal series is typically the next step in management, although if volvulus is suspected, caution with non water soluble contrast is mandatory as the usage of barium can impede surgical revision and lead to increased post operative complications."*

Matched Pattern:

Semantic Hierarchy (after the first transformation pass):

(2b) *"although"* → *Contrast*

(2a) core   **(2a)** core

(1)   **(1)**
A fluoroscopic study   If volvulus is suspected,
which is known as an upper   caution with non water soluble contrast
gastrointestinal series   is mandatory as the usage of barium
is typically the next   can impede surgical revision and lead
step in management.   to increased post operative complications.

Figure 4.2 (Subtask 1) The source sentence is split up and rephrased into a set of syntactically simplified sentences. **(Subtask 2a)** Then, the split sentences are connected with information about their constituency type to establish a contextual hierarchy between them. **(Subtask 2b)** Finally, by identifying and classifying the rhetorical relation that holds between the simplified sentences, their semantic relationship is captured, which can be used to inform downstream Open IE applications

The transformation rules encode both the *splitting points* and the *rephrasing procedures* for reconstructing grammatically sound sentences. Each rule takes a sentence's phrasal parse tree[10] as input and encodes a pattern that, in case of a match, will extract specified constituents from the tree. The decomposed text spans, as well as the remaining components are then transformed into syntactically well-formed stand-alone sentences. For instance, in order to ensure that the resulting simplified output is grammatically sound, the extracted spans are combined with their corresponding referents from the main sentence or appended to a simple canonical phrase (e.g., *"This is"*), if necessary.

For a better understanding of the splitting procedure, Figure 4.2 visualizes the application of the first grammar rule that matches the input sentence from Table 1.1. The upper part of the box represents the complex source sentence, whose phrasal parse tree is matched against the specified simplification pattern (see Table 6.2a). The lower part then depicts the result of the corresponding transformation operation. The remaining steps, subtasks 2a and 2b, are detailed below in Section 5.

4.3 Execution Order of the Transformation Patterns

The grammar rules are applied recursively in a top-down fashion. When no more rule matches the set of simplified sentences, the algorithm terminates. The transformation patterns are executed in a fixed order that was empirically determined by examining which sequence achieved the best simplification results. For this purpose, we conducted a manual qualitative analysis on a development set of 300 sentences from Wikipedia that were sampled for heterogeneity to ensure that they present a diverse structure.[11]

With respect to the order of execution, the 35 specified grammar rules can be grouped into 17 classes that are applied in the order shown in Table 4.3.[12]

The execution order of the transformation patterns was balanced for various criteria, including the frequency of the rules, their granularity, complexity and specificity, as well as susceptibility to errors. In general, the least error-prone rules are executed first (e.g., in a manual analysis, we determined that 99.1% of the coordinate clauses (1) were correctly split (see Section 15.1.2.2)). During the transformation process,

---

[10] generated by Stanford's pre-trained lexicalized parser (Socher et al., 2013)

[11] To demonstrate diversity in the syntactic patterns of the sampled sentences, a detailed study of the distribution of the different clausal and phrasal constructs over this dataset was carried out. For details, see Section 19.3 in the online supplemental material.

[12] The exact execution sequence of the individual transformation rules can be found in Section 19.1 in the online supplemental material.

Table 4.3 Execution order of the transformation patterns

ORDER	RULE GROUP
1	coordinate clauses
2	non-restrictive relative clauses
3	appositive phrases
4	preposed adverbial clauses and participial phrases
5	coordinate verb phrases
6	postposed adverbial clauses and participial phrases
7	reported speech with preposed attribution
8	postposed adverbial clauses
9	reported speech with postposed attribution
10	embedded participial phrases
11	restrictive relative clauses
12	prepositional phrases that act as complements of verb phrases
13	postposed participial phrases
14	adjectival/adverbial phrases
15	lead noun phrases
16	prepositional phrases that are offset by commas
17	coordinate noun phrase lists

we then work our way to the more error-prone ones, such as the rules for decomposing lists of noun phrases (17), which can easily be confused with appositive phrases. Beyond that, the patterns become more and more complex towards the end of the list, as we need to check for special cases and possibilities of confusion (e.g., (12) and (17)). It also becomes clear that we first operate on coarse-grained level, dividing up clauses (1-11), before we go down to the phrasal level (10-17), resulting in much more fine-grained splits. Moreover, specific rules are carried out early in the process, such as the rules for disembedding relative clauses (lexicalized on the relative pronouns, 2) and appositive phrases (based on named entities, 3). More general rules, on the other hand, tend to be executed later. Finally, there is a rough orientation on the frequency of the rules, starting with the ones that are triggered more often, and working towards the ones that occur less frequently in the transformation process (see Section 15.1.1.4).

To sum up, the order of execution of the transformation patterns is based on the following five criteria, which have been balanced against each other:

(1) **susceptibility to errors**: from less to more error-prone
(2) **granularity**: from clausal to phrasal units
(3) **complexity**: from simple to complex
(4) **specificity**: from specific to general
(5) **frequency**: from frequent to infrequent

Subtask 2: Establishing a Semantic Hierarchy

5

Each split will create two or more sentences with a simplified syntax. In order to establish a semantic hierarchy between them, two subtasks are carried out: constituency type classification and rhetorical relation identification (see Figure 5.1).

SUBTASK 2: ESTABLISH A SEMANTIC HIERARCHY

Input: A pair of structurally simplified sentences T_i and T_j.

Problem: Establish a *contextual hierarchy* between the split sentences and capture their *semantic relationship.*

Figure 5.1 Problem description of subtask 2, whose goal is to establish a semantic hierarchy between the simplified sentences

5.1 Constituency Type Classification

To preserve the coherence of the decomposed spans, we set up a contextual hierarchy between the split sentences in a first step. For this purpose, we connect them with information about their constituency type, signifying the relative importance of the content expressed in the respective utterances: two spans of equal importance are both referred to as "core sentences"; the same applies to units that reflect the main message of the input. On the contrary, utterances that provide some piece of background information are qualified as "context sentences" in our approach.

Supplementary Information The online version contains supplementary material available at https://doi.org/10.1007/978-3-658-38697-9_5.

C. Niklaus, *From Complex Sentences to a Formal Semantic Representation using Syntactic Text Simplification and Open Information Extraction*,
https://doi.org/10.1007/978-3-658-38697-9_5

According to Fay (1990), clauses can be related to one another in two ways: first, there are parallel clauses that are linked by coordinating conjunctions, and second, clauses may be embedded inside another, introduced by subordinating conjunctions. The same is true for phrasal elements. Since subordinations commonly express less relevant information, we denote them "context sentences". In contrast, coordinations are of equal status and typically depict the key information contained in the input. Therefore, they are called "core sentences" in our approach. To differentiate between those two types of constituents, the transformation patterns encode a simple syntax-based method where subordinate clauses and subordinate phrasal elements are classified as context sentences, while their superordinate counterparts as well as coordinate clauses and coordinate phrases are labelled as core (see Table 4.2).

However, two exceptions have to be noted. In case of an attribution (e.g. *"Obama announced that he would resign his Senate seat."*), the subordinate clause expressing the actual statement (*"He would resign his Senate seat."*) is assigned a core tag, while the superordinate clause containing the fact that it was uttered by some entity (*"This was what Obama announced."*) is labelled as contextual information. The reason for this is that the latter is considered less relevant as compared to the former, which holds the key information of the source sentence. Moreover, when disembedding adverbial clauses of contrast (as determined in the rhetorical relation identification step, see Section 5.2), both the superordinate and the subordinate clause are labelled

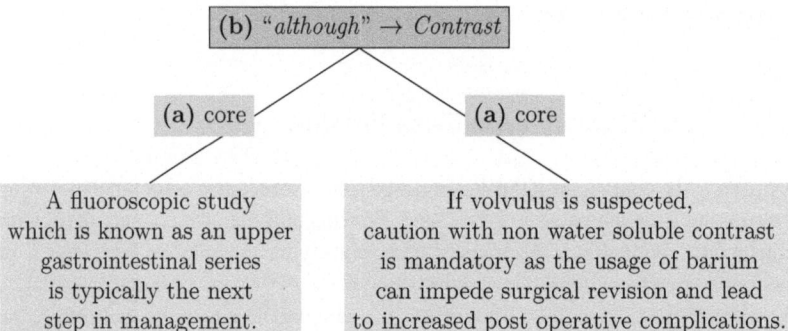

Figure 5.2 Result of the constituency type classification and rhetorical relation identification steps after the first transformation pass on our running example from Table 1.1. In subtask (a), the split sentences are connected with information about their constituency type to establish a contextual hierarchy between them. Next, in subtask (b), the rhetorical relation that holds between the simplified sentences is identified and classified in order to preserve their semantic relationship. In that way, a semantic hierarchy is established between the decomposed spans

as a core sentence (see step (a) in the example in Figure 5.2), in accordance with the theory of RST, where text spans that present a "Contrast" relationship are considered both as a nucleus.

This approach allows for the distinction of core information capturing the key message of the input from contextual information that provides only supplementary material, resulting in a two-layered hierarchical representation in the form of core facts and accompanying contextual information. This directly relates to the concept of nuclearity in RST, which specifies each text span as either a nucleus or a satellite. The nucleus span embodies the central piece of information and is comparable to what we denote a core sentence, whereas the role of the satellite is to further specify the nucleus, corresponding to a context sentence in our case.

5.2   Rhetorical Relation Identification

In addition to capturing the hierarchical order between the decomposed spans, we aim to preserve their semantic context. For this purpose, we identify and classify the rhetorical relationship that holds between the simplified sentences, making use of both syntactic and lexical features, which we encode in the transformation patterns.

While syntactic features are manifested in the phrasal composition of a sentence's parse tree, lexical features are extracted from the parse tree in the form of signal spans that indicate cue phrases for the identification of rhetorical relations. Based on the work of Knott and Dale (1994), we determine potential signal spans to extract, as well as their positions in specific syntactic environments, and encode them as features in the transformation rules. Each signal span represents a sequence of words that are likely to indicate a semantic relationship between two spans. To infer the type of rhetorical relation that is signified by a given signal span, we make use of a predefined list of rhetorical cue words adapted from the work of Taboada and Das (2013). The authors analyze how different rhetorical relations are signalled in texts, demonstrating that the identification of rhetorical relations is predominantly based on lexical cue phrases (e.g., *Background, Cause, Condition, Contrast*) and syntactic structures (e.g., *Purpose* and *Attribution*). Based on their findings, they define a list of cue phrases and map each of them to the rhetorical relation that they most likely trigger.[1] Consequently, if the signal span that is extracted according to the respective transformation pattern corresponds to one of the cue phrases specified

---

[1] The full list of cue phrases that serve as lexical features for the identification of rhetorical relations in our discourse-aware TS approach, as well as the corresponding relations to which they are mapped, is provided in the online supplemental material in Section 20.1.

in this list, the associated rhetorical relation is established between the decomposed utterances. Otherwise, a default relation is set, i.e. *"Unknown-Coordination"* in case of coordinate spans or *"Unknown-Subordination"* if one of the split components is subordinate to the other. For example, the transformation rule in Table 6.2a, which is the first pattern that matches our running example, specifies that the phrase *"although"* is the cue word here, which is mapped to a "Contrast" relationship according to the findings in Taboada and Das (2013) (see step (b) in the example in Figure 5.2). Beyond that, some of the patterns are explicitly tailored to identify a small set of selected rhetorical relations that heavily rely on syntactic features, including the *"Purpose"* and *"Attribution"* relationships. In this case, no signal span is extracted.

The set of rhetorical relations that we employ for this task is listed in Table 5.1. They constitute a subset of the classical set of RST relations defined in Mann and Thompson (1988). Additionally, we adopt a number of relations from the extended set of rhetorical relations defined in the RST-DT, including *Temporal-After*, *Temporal-Before*, *Sequence*, *Inverted-Sequence* and *Attribution*. In order to deal with the problem of linking split sentences whose connecting rhetorical relation could not be identified, the custom relations *Unknown-Coordination* and *Unknown-Subordination* are introduced. Moreover, representing classes of context that were frequently encountered in our example sentences, we add the relations *Spatial* and *Temporal* for the identification of semantic relationships that provide local or temporal information about a presented situation. Their definitions are provided in Table 5.2.

The process of selecting the rhetorical relations to include in our approach was guided by the following two questions (Cetto, 2017):

- Which rhetorical relations are the most relevant for downstream Open IE applications?
- Which of them are likely to be recognized by syntactic and lexical features, according to the work of Taboada and Das (2013)?

Furthermore, some adaptations had to be made (Cetto, 2017). Since in our approach, the rhetorical relations are used to semantically classify the contextual sentences, we need to ensure that the context span (corresponding to the satellite span in RST) is the unit that is characterized by the relation name. However, this differs from some of the definitions found in the literature. For instance, the "Cause", "Result", "Temporal-After" and "Temporal-Before" relations in the RST-DT refer to the situation that is presented in the nucleus (Carlson and Marcu, 2001). Consequently, in these cases, the relationship has to be inverted. Moreover, for coordinate relations

Table 5.1 Set of rhetorical relations employed for the rhetorical relation identification step (Cetto, 2017)

RHETORICAL RELATION / INVERSE RELATION	CORE SPAN	FOLLOWING CORE SPAN
Unknown-Coordination / Unknown-Coordination	a syntactically coordinate span (default)	a syntactically coordinate span (default)
Contrast / Contrast	one alternate	the other alternate
Cause / Result	a situation	another situation which causes that one
Result / Cause	a situation	another situation which is caused by that one
List / List	an item	a next item
Disjunction / Disjunction	an item	an alternate item
Temporal-After / Temporal-Before (Sequence)	a situation	a situation that occurs after that
Temporal-Before / Temporal-After (Inverted-Sequence)	a situation	a situation that occurs before that

(a) Coordinations.

RHETORICAL RELATION	CORE SPAN	CONTEXT SPAN
Unknown-Subordination	the syntactically superordinate span (default)	the syntactically subordinate span (default)
Attribution	the reported message	the source of the attribution
Background	text whose understanding is being facilitated	text for facilitating understanding
Cause	a situation	another situation which causes that one
Result	a situation	another situation which is caused by that one
Condition	action or situation whose occurrence results from the occurrence of the conditioning situation	conditioning situation
Elaboration	basic information	additional information
Purpose	an intended action	the intent behind the situation
Temporal-After	a situation	a situation that occurs after that
Temporal-Before	a situation	a situation that occurs before that

(b) Subordinations.

Table 5.2 Set of additional contextual relations (Cetto, 2017)

RELATION	CORE SPAN	CONTEXT SPAN
Spatial	a situation	spatial information that describes where the situation took place
Temporal	a situation	temporal information that describes when the situation happened

(corresponding to multi-nuclear relations in RST), the relation name specifies the span on the right position, according to the theory of RST. Hence, in order to allow for a classification of both spans in the case of a coordination, we introduced inverse relations, where the core span to the right is classified with the original relation from the definition, whereas its left counterpart is flagged with the corresponding inverse relation (see Table 5.1a). Note that "Temporal-After" and "Temporal-Before" relations are defined for both coordinations and subordinations. In the former case, both spans are considered equally important, whereas in the latter case, the context span is deemed less relevant. To be exact, the definitions of the coordinate "Temporal-After" and "Temporal-Before" relations correspond to that of the "Sequence" and "Inverted-Sequence" relations in the RST-DT.

Figure 5.2 depicts the result of the first transformation pass on the example sentence from Table 1.1, establishing a semantic hierarchy between the decomposed sentence pair. A detailed step-by-step example demonstrating the complete transformation process on this sentence is provided in Chapter 7. Prior to that, we will describe in detail the set of transformation patterns that we specified in order to split complex sentences into a set of hierarchically ordered and semantically interconnected minimal propositions.

Transformation Patterns 6

During the transformation process described in Chapters 4 and 5, we operate both on the clausal and phrasal level of a sentence by recursively

- *splitting and rephrasing complex multi-clause sentences* into sequences of simple sentences that each contain exactly one independent clause, and
- *extracting selected phrasal components* into stand-alone sentences (see Section 4.2).

By connecting the decomposed sentences with information about their constituency type, a two-layered hierarchical representation in the form of core sentences and accompanying contexts is created (see Section 5.1). Finally, in order to preserve their semantic relationship, the rhetorical relations that hold between the split components are identified and classified (see Section 5.2). In that way, complex input sentences are transformed into a semantic hierarchy of minimal propositions that present a simple and regular structure (see Figures 1.2 and 1.3). Thus, we create an intermediate representation which is targeted at supporting machine processing in downstream Open IE tasks whose predictive quality deteriorates with sentence length and structural complexity (see Section 15.3). In the following, we will discuss in detail the 35 patterns that we specified to carry out this transformation process.

Supplementary Information The online version contains supplementary material available at https://doi.org/10.1007/978-3-658-38697-9_6.

6.1   Clausal Disembedding

Fay (1990) distinguishes four central functional categories of clauses which can be divided into two combination methods: *coordination* and *subordination*. A **coordinate clause** is a clause that is connected to another clause with a coordinating conjunction, such as *"and"* or *"but"*. In contrast, a subordinate clause is a clause that begins with a subordinating conjunction, e.g. *"because"* or *"while"*, and which must be used together with a main clause. The class of subordinate clauses can be subdivided into the following three groups:

(1) **adverbial clauses,**
(2) **relative clauses** and
(3) **reported speech.**

All four clause types are handled by our proposed discourse-aware128 TS approach, as detailed below.

6.1.1   Coordinate Clauses

Coordinate clauses are two or more clauses in a sentence that have the same status and are joined by coordinating conjunctions, e.g. *"and"*, *"or"* or *"but"* (Fay, 1990). When given such a syntactic structure, each clause is extracted from the input and transformed into a separate simplified sentence according to the rule depicted in Figure 6.1. To avoid important contextual references from being lost (that may be contained, for example, in prepositional phrases at the beginning or end of the source sentence), each of the separated clauses is concatenated with the textual spans z_1 and z_2 that precede or follow, respectively, the coordinate clauses. In the event of a pair of coordinate clauses, the span that connects them acts as the signal span (x) which is mapped to its corresponding rhetorical relation to infer the type of semantic relationship that holds between the decomposed elements (Cetto, 2017). Since coordinate clauses commonly are of equal importance, each simplified output is labelled as a core sentence.

The simplification rule that defines how to split up and transform coordinate clauses is specified in terms of Tregex patterns[1] in Table 6.1. For a better understanding, Figure 6.1 illustrates this rule in the form of a phrasal parse tree, and Figure 6.2 shows how an example sentence is mapped to this pattern and converted

---

[1] For details on the rule syntax, see Levy and Andrew (2006).

Table 6.1 The transformation rule pattern for splitting coordinate clauses (RULE #1). The pattern in bold represents the part of a sentence that is extracted from the source and turned into a stand-alone sentence. The underlined pattern (solid line) will be deleted from the remaining part of the input. The pattern that is underlined with a dotted line serves as a cue phrase for the rhetorical relation identification step

TREGEX PATTERN	ROOT <<: (S < (S ?$.. CC & $.. S))
EXTRACTED SENTENCE	**S.**
EXAMPLE INPUT	Many consider the flavor to be very agreeable, but **it is generally bitter if steeped in boiling water.**
SEMANTIC HIERARCHY	

COORD
Contrast

core   core

Many consider the   It is generally
flavor to be   bitter if steeped
very agreeable.   in boiling water.

into a semantic hierarchy by (1) breaking up the input into a sequence of simplified sentences, (2) setting up a contextual hierarchy between the split components, and (3) identifying the semantic relationship that holds between them.[2]

6.1.2 Adverbial Clauses

One major category of subordinate clauses are adverbial clauses. An adverbial clause is a dependent clause that functions as an adverb, adding information that describes how an action took place (e.g., when, where and how). It is fronted by a subordinating conjunction, such as *"since"*, *"while"*, *"after"*, *"if"* or *"because"*. Depending on the actions or senses of their conjunctions, adverbial clauses are grouped into different classes, including *time, place, condition, contrast* and *reason* (Quirk et al., 1985).

---

[2] In the discussion of the remaining linguistic constructs, we will restrict ourselves to reporting the Tregex patterns that we specified for carrying out the transformation process. Besides, some selected rules will be illustrated in the form of a phrasal parse tree and their application to an example sentence in the online supplemental material (see Section 19.2).

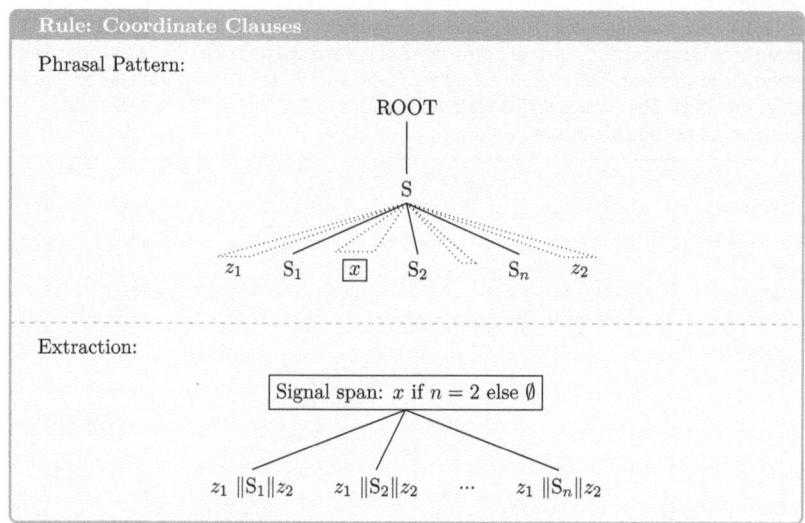

Figure 6.1 Rule for splitting coordinate clauses (Cetto, 2017)

Sentences containing an adverbial clause are split into two simplified components. One of them corresponds to the superordinate statement, whereas the other embodies the subordinate clause. Syntactic variations of such structures cover the following linguistic expressions, differing in the order of their clausal components:

(1) the subordinate clause follows the superordinate span;
(2) the subordinate clause precedes the superordinate span; or
(3) the subordinate clause is positioned between discontinuous parts of the superordinate span.

The subordinate clause is always introduced with a discourse connective (Prasad et al., 2007). It serves as a cue phrase that is mapped to its corresponding rhetorical relation in order to infer the type of semantic relationship that holds between the decomposed elements (Cetto, 2017).[3] Since the sentence originating from the adverbial clause typically presents a piece of background information, it is marked with a context label, while the one corresponding to its superordinate span is tagged

---

[3] RULE #4 to #7 are lexicalized on the preposition *"to"* and always mapped to a "Purpose" relationship.

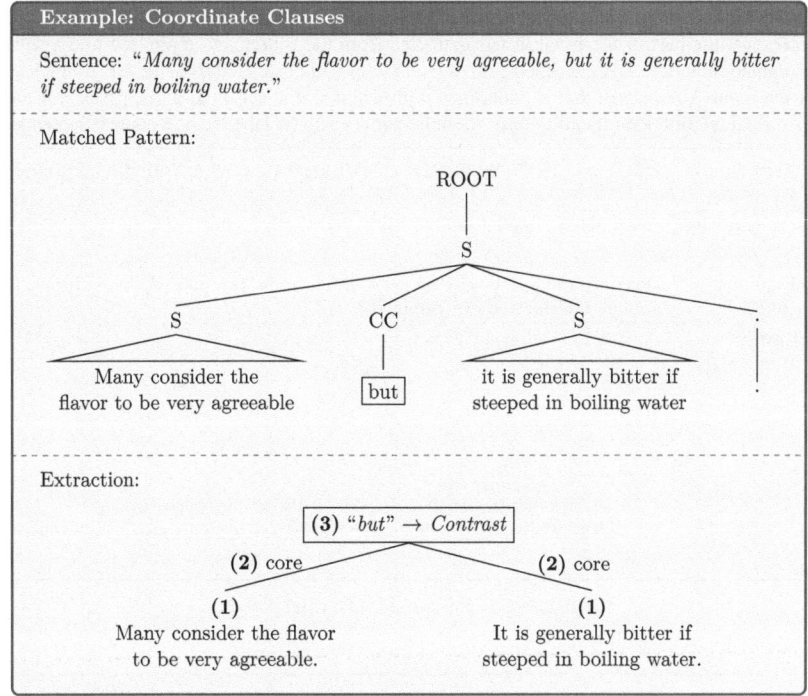

Figure 6.2 Example for splitting coordinate clauses (RULE #1)

as a core sentence. As an example, the implementation of a rule pattern where the subordinate clause follows the superordinate span is shown in Figure 19.1 in the online supplemental material. Figure 19.2 in the online supplemental material then illustrates its application to an example sentence. The full set of Tregex patterns that were specified in order to transform sentences containing adverbial clauses into a simplified hierarchical representation are displayed in Table 6.2.

6.1.3   Relative Clauses

A relative clause is a clause that is attached to its antecedent by a relative pronoun, such as *"who"*, *"which"* or *"where"*. There are two types of relative clauses, differing in the semantic relation between the clause and the phrase to which it

Table 6.2 The transformation rule patterns addressing adverbial clauses. A pattern in bold represents the part of the input that is extracted from the source and rephrased into a self-contained sentence. An underlined pattern (solid line) will be deleted from the remaining part of the input. The pattern that is underlined with a dotted line serves as a cue phrase for the rhetorical relation identification step. An italic pattern will be labelled as a context sentence

TREGEX PATTERN	ROOT <<: (S < (NP $.. (VP < +(VP) (<u>SBAR</u> < (*S* < (*NP $.. VP*))))))
EXTRACTED SENTENCE	*S* < (*NP $.. VP*).
EXAMPLE INPUT	Donald Trump was elected over Democratic nominee Hillary Clinton, although **he lost the popular vote.**
SEMANTIC HIERARCHY	

(a) RULE #2: Postposed adverbial clauses.

TREGEX PATTERN	ROOT <<: (S < (<u>SBAR</u> < (*S* < (*NP $.. VP*)) $.. (*NP $.. VP*)))
EXTRACTED SENTENCE	*S* < (*NP $.. VP*).
EXAMPLE INPUT	Though **shorts are an option for many casual occasions,** they may also be inappropriate for more formal occasions.
SEMANTIC HIERARCHY	

(b) RULE #3: Preposed adverbial clauses.

(Continued)

Table 6.2 (Continued)

TREGEX PATTERN	ROOT <<: (S < (NP $.. (VP < +(VP) (NP|PP $.. *(S* <<, *(VP* <<, *//(T|t)o/))))))*
EXTRACTED SENTENCE	*"This"* + BE + *S* <<, *(VP* <<, */(T|t)o/).*
EXAMPLE INPUT	He was sent to Geneva in 1929 *to act as Ireland's representative to the League of Nations.*
SEMANTIC HIERARCHY	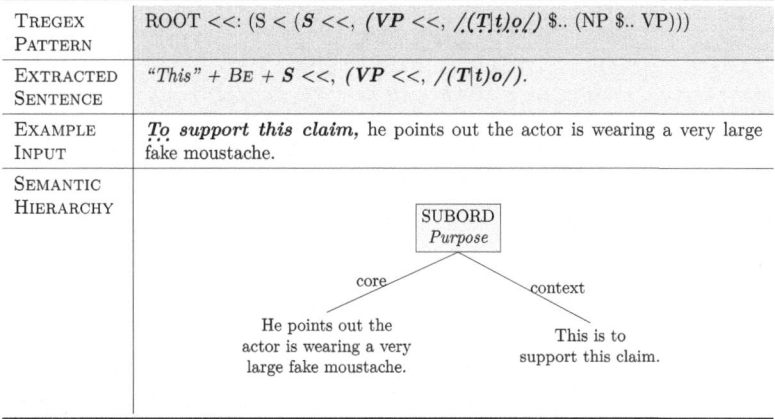

(c) RULE #4: Postposed adverbial clauses of purpose introduced by the phrase *"to do"*.

TREGEX PATTERN	ROOT <<: (S < (*S* <<, *(VP* <<, */(T|t)o/)* $.. (NP $.. VP)))
EXTRACTED SENTENCE	*"This"* + BE + *S* <<, *(VP* <<, */(T|t)o/).*
EXAMPLE INPUT	*To support this claim,* he points out the actor is wearing a very large fake moustache.
SEMANTIC HIERARCHY	SUBORD Purpose — core: He points out the actor is wearing a very large fake moustache. — context: This is to support this claim.

(d) RULE #5: Preposed adverbial clauses of purpose introduced by the phrase *"to do"*.

(Continued)

Table 6.2 (Continued)

TREGEX PATTERN	ROOT <<: (S < (NP $.. (VP < +(VP) (*SBAR* < *(S <<, (VP <<, /(T|t)o/))))))*
EXTRACTED SENTENCE	*"This"* + BE + *SBAR* < *(S <<, (VP <<, /(T|t)o/)).*
EXAMPLE INPUT	A graduate student is researching the evolution of human eyes, *in order to discredit creationists by proving that eyes have evolved.*
SEMANTIC HIERARCHY	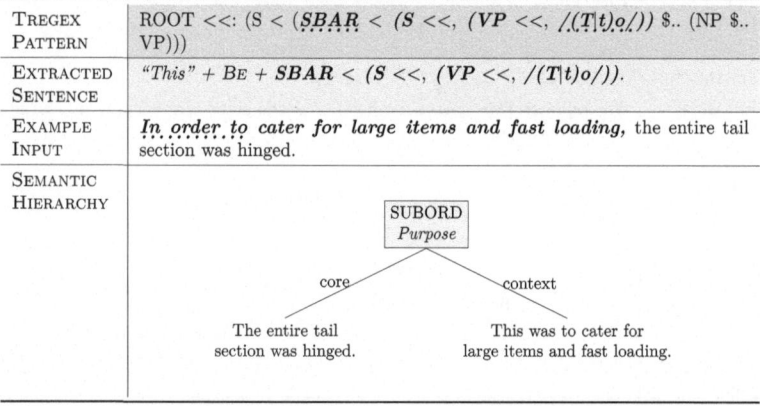

(e) RULE #6: Postposed adverbial clauses of purpose introduced by the phrase *"in order to do"*.

TREGEX PATTERN	ROOT <<: (S < (*SBAR* < *(S <<, (VP <<, /(T|t)o/))* $.. (NP $.. VP)))
EXTRACTED SENTENCE	*"This"* + BE + *SBAR* < *(S <<, (VP <<, /(T|t)o/)).*
EXAMPLE INPUT	*In order to cater for large items and fast loading,* the entire tail section was hinged.
SEMANTIC HIERARCHY	

SUBORD
Purpose

core   context

The entire tail   This was to cater for
section was hinged.   large items and fast loading.

(f) RULE #7: Preposed adverbial clauses of purpose introduced by the phrase *"in order to do"*.

Table 6.3 The transformation rule patterns addressing non-restrictive relative clauses. A pattern in bold represents the part of the input that is extracted from the source and rephrased into a self-contained sentence. A boxed pattern refers to its referent. An underlined pattern (solid line) will be deleted from the remaining part of the input. An italic pattern will be labelled as a context sentence

TREGEX PATTERN	ROOT <<: (S << (NP <, NP & < (/,/ \$+ (SBAR <, (WHPP \$+ *S* & <, *IN* & < – WHNP) & ?\$+ /,/))))
EXTRACTED SENTENCE	*S + IN +* NP.
EXAMPLE INPUT	He had been enrolled into Harvard University , *at which* *he studied archaeology.*
SEMANTIC HIERARCHY	

(a) RULE #8: Non-restrictive relative clauses commencing with a preposition followed by a relative pronoun.

TREGEX PATTERN	ROOT <<: (S << (/.*/ < (NP|PP \$+ (/,/ \$+ (SBAR <, (WHADVP \$+ *S* & <<: WRB) & ?\$+ /,/)))))
EXTRACTED SENTENCE	*S.*
EXAMPLE INPUT	The noted communist, Sakhavu Kurumpakara Thankappan, was born and raised in Kurumpakara, where *a memorial dedicated to him is situated at the Udayonmuttam Junction.*
SEMANTIC HIERARCHY	

(b) RULE #9: Non-restrictive relative clauses commencing with the relative pronoun "where".

(Continued)

Table 6.3 (Continued)

TREGEX PATTERN	ROOT <<: (S << (NP <, NP & < (/,/ $+ (SBAR <, (WHNP $+ (*S* <, *NP=np_rel* & < − *(VP=vp_rel* ? < *+(VP) PP))* & <<: (WP <: whom)) & ?$+ /,/))))
EXTRACTED SENTENCE	*np_rel* + *vp_rel* + NP + *PP*.
EXAMPLE INPUT	He is best known for his work with The Byrds , whom **he joined in September 1968.**
SEMANTIC HIERARCHY	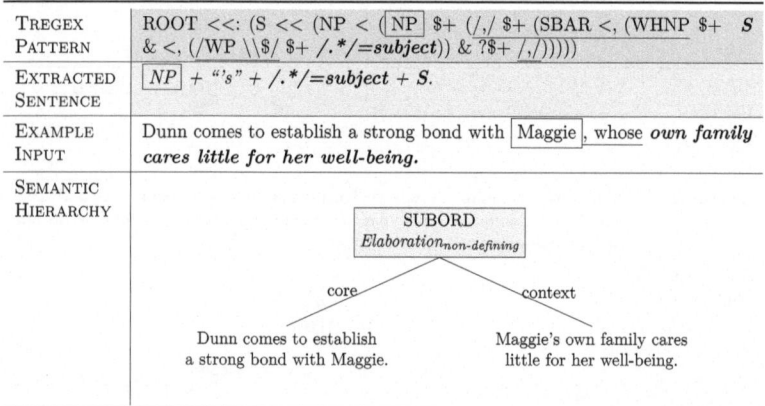

(c) RULE #10: Non-restrictive relative clauses commencing with the relative pronoun "whom".

TREGEX PATTERN	ROOT <<: (S << (NP < (NP $+ (/,/ $+ (SBAR <, (WHNP $+ *S* & <, (/WP \\$/ $+ /.*/=subject)) & ?$+ /,/)))))
EXTRACTED SENTENCE	NP + *"'s"* + /.*/=subject + *S*.
EXAMPLE INPUT	Dunn comes to establish a strong bond with Maggie , whose **own family cares little for her well-being.**
SEMANTIC HIERARCHY	SUBORD *Elaboration_{non-defining}* core / context / Dunn comes to establish a strong bond with Maggie. / Maggie's own family cares little for her well-being.

(d) RULE #11: Non-restrictive relative clauses commencing with the relative pronoun "whose".

(Continued)

Table 6.3 (Continued)

TREGEX PATTERN	ROOT <<: (S << (NP <, $\boxed{\text{NP}}$ & < (/,/ $+ (SBAR <, (WHNP $+ ***S*** & <<: WP\|WDT) & ?$+ /,/))))
EXTRACTED SENTENCE	$\boxed{\text{NP}}$ + ***S***.
EXAMPLE INPUT	She met $\boxed{\text{her husband}}$, <u>who</u> ***was completing his doctorate in physics.***
SEMANTIC HIERARCHY	

SUBORD

Elaboration$_{non\text{-}defining}$

core   context

She met her   Her husband was completing
husband.   his doctorate in physics.

(e) RULE #12: Non-restrictive relative clauses commencing with either the relative pronoun *"who"* or *"which"*.

Table 6.4 The transformation rule patterns addressing restrictive relative clauses. A pattern in bold represents the part of the input that is extracted from the source and rephrased into a self-contained sentence. A boxed pattern refers to its referent. An underlined pattern (solid line) will be deleted from the remaining part of the input. An italic pattern will be labelled as a context sentence

TREGEX PATTERN	ROOT <<: (S << (NP <, $\boxed{\text{NP}}$ & < (SBAR <, (WHNP $+ ***(S <, NP &*** *< − (VP ? < +(VP) PP))* & <<: $\underline{\text{(WP <: whom}}$)))))
EXTRACTED SENTENCE	***(S <, NP & < − (VP ? < +(VP) PP))*** + $\boxed{\text{NP}}$.
EXAMPLE INPUT	$\boxed{\text{The artist}}$ <u>whom</u> ***she admires*** won an award.
SEMANTIC HIERARCHY	

SUBORD

Elaboration$_{defining}$

core   context

The artist won   She admires
an award.   the artist.

(a) RULE #13: Restrictive relative clauses commencing with the relative pronoun *"whom"*.

(Continued)

Table 6.4 (Continued)

Tregex Pattern	ROOT <<: (S << (NP < (NP $+ (SBAR <, (WHNP $+ *S* & <, (/ WP\\$/ $+ /.*/=subject))))))
Extracted Sentence	NP + "'s" + *subject* + *S*.
Example Input	The rescue operation to reach Flight 608 was carried out by the Canadian Forces whose *plane spotted the downed aircraft.*
Semantic Hierarchy	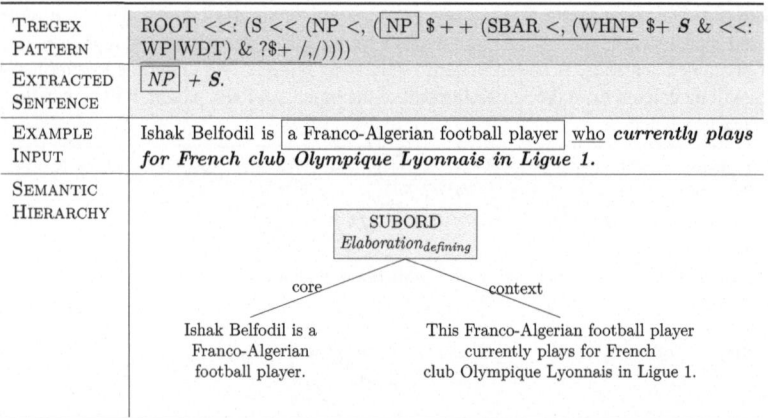

(b) RULE #14: Restrictive relative clauses commencing with the relative pronoun *"whose"*.

Tregex Pattern	ROOT <<: (S << (NP <, (NP $+ + (SBAR <, (WHNP $+ *S* & <<: WP\|WDT) & ?$+ /,/))))
Extracted Sentence	NP + *S*.
Example Input	Ishak Belfodil is a Franco-Algerian football player who *currently plays for French club Olympique Lyonnais in Ligue 1.*
Semantic Hierarchy	<table continues below>

SUBORD
Elaboration_{defining}

core — context

Ishak Belfodil is a
Franco-Algerian
football player.

This Franco-Algerian football player
currently plays for French
club Olympique Lyonnais in Ligue 1.

(c) RULE #15: Restrictive relative clauses commencing with either the relative pronoun *"who"*, *"which"* or *"that"*.

(Continued)

Table 6.4 (Continued)

TREGEX PATTERN	ROOT <<: (S << (NP <, (NP $ + + (SBAR <: *(S < (VP ? < (PP ? <: IN)))))))*
EXTRACTED SENTENCE	*(S < (VP ? < (PP ? <: IN)))* + NP .
EXAMPLE INPUT	The novelist *she adores* published a new book.
SEMANTIC HIERARCHY	

```
            SUBORD
        Elaboration_defining

    core              context

The novelist published    She adores
   a new book.            the novelist.
```

(d) RULE #16: Reduced relative clauses, which are not marked by an explicit relative pronoun.

refers: restrictive and non-restrictive. In the former case, the relative clause is strongly connected to its antecedent, providing information that identifies the phrasal component it modifies (e.g., *"Obama criticized **leaders** who refuse to step off."*) Hence, it supplies essential information and therefore cannot be eliminated without affecting the meaning of the sentence. In contrast, non-restrictive relative clauses are parenthetic comments which describe, but do not further define their antecedent (e.g., *"Obama brought attention to the **New York City Subway System**, which was in a bad condition at the time ."*), and thus can be left out without disrupting the meaning or structure of the sentence (Quirk et al., 1985). As non-restrictive relative clauses are set off by commas, unlike their restrictive counterparts, they can be easily distinguished from one another on a purely syntactic basis.

In order to identify whether a given sentence includes a non-restrictive relative clause, we check if one of the Tregex patterns of Table 6.3 matches its phrasal parse tree. If so, a self-contained context sentence providing additional information about the referred phrase is constructed by linking the relative clause (without the relative pronoun) to the phrase that has been identified as its antecedent. At the same time, the source sentence is reduced to its key information by dropping the extracted relative clause, and tagged as a core sentence. In this way, subordinations that are introduced by one of the relative pronouns *"who"*, *"whom"*, *"whose"*, *"which"* or *"where"*, as well as a combination of a preposition and one of the pronouns mentioned before are handled by our approach. The example in Figure 19.4 in the online supplemental material illustrates this procedure. It matches the transformation pattern from Figure 19.3 in the online supplemental material, which shows - in terms of a phrasal parse tree - the rule for treating non-restrictive relative clauses commencing with either the relative pronoun *"who"* or *"which"*.

As opposed to non-restrictive relative clauses, their restrictive counterparts represent an integral part of the phrase to which they are linked. Regardless, according to our goal of splitting complex input sentences into minimal propositions, we decided to decompose the restrictive version of relative clauses as well, and transform them into stand-alone context sentences according to the rules listed in Table 6.4. To clarify that the detached components provide irremissible information about their referent, we insert the adjunct "defining" when we add the semantic relationship between the split sentences in the form of a rhetorical relation. Both types of relative clauses establish an "Elaboration" relation, since they present additional detail about the entity to which they refer.[4] For the sake of clarity, we append the adjunct "non-defining" in case of a non-restrictive relative clause.

6.1.4   Reported Speech

Special attention was paid to attribution relationships expressed in reported speech, e.g. *"Obama announced [that he would resign his Senate seat.]$_{SBAR}$*, which, too, fall into the syntactic category of subordinations. To distinguish this type of linguistic expression from adverbial clauses, we defined a set of additional rule patterns that target the detection of attributions (Cetto, 2017). Similar to Mausam et al. (2012), we identify attributions by matching the lemmatized version of the head verb of the sentence (here: *"announce"*) against a list of verbs of reported speech and cognition (Carlson and Marcu, 2001).[5]

The Tregex patterns for splitting up and turning reported speech constructs into a two-layered hierarchy of simplified sentences are listed in Table 6.5. The decomposed parts are always connected via an "Attribution" relation. Moreover, note that there is a peculiarity here with regard to the contextual hierarchy. As opposed to previously mentioned cases, where subordinations are marked up as context sentences, while the corresponding superordinate spans are labelled as core sentences,

---

[4] The only exception are relative clauses commencing with the relative pronoun *"where"* (RULE #9). As they provide details about a location, we assign a "Spatial" relationship between the split sentences in this case.

[5] The full list of verbs that we specified for identifying attributions can be found in Section 20.2. in the online supplemental material.

Table 6.5 The transformation rule patterns addressing reported speech constructs. A pattern in bold represents the part of the input that is extracted from the source and rephrased into a self-contained sentence. An underlined pattern (solid line) will be deleted from the remaining part of the input. A pattern that is underlined with a dotted line serves as a cue phrase for the rhetorical relation identification step. An italic pattern will be labelled as a context sentence

Tregex Pattern	ROOT <<: (S < (S|SBAR|SBARQ $.. *(NP [$,, VP=vp | $.. VP=vp]))*); *vb* is an attribution-verb
Extracted Sentence	*"This" + BE + "what" + (NP [$,, VP=vp | $.. VP=vp]).*
Example Input	Witness memories don't get better with time, *she said in an interview with the International Herald Tribune.*
Semantic Hierarchy	

SUBORD
Attribution

core   context

Witness memories don't   This is what she said
get better with time.   in an interview with the
                       International Herald Tribune.

(a) Rule #17: Reported speech with postposed attribution.

Tregex Pattern	ROOT <<: (S < (NP $.. (VP < +(VP)) (SBAR [,, /"/=start] <<, /"/=start] [.. /"/=end] << – /"/=end]))))
Extracted Sentence	*"This" + BE + "what" + S < (NP $.. (VP < +(VP)).*
Example Input	*Pauli remarked sadly* "It is not even wrong".
Semantic Hierarchy	

SUBORD
Attribution

core   context

"It is not even wrong."   This was what
                   Pauli remarked sadly.

(b) Rule #18: Direct speech with preposed attribution.

(Continued)

Table 6.5 (Continued)

TREGEX PATTERN	ROOT <<: (S < (S|SBAR|SBARQ [.. /"/=start | <<, /"/=start] [.. /"/=end | << − /"/=end] $.. *(NP [$,, VP | $.. VP]))*
EXTRACTED SENTENCE	*"This" + BE + "what" + (NP [$,, VP | $.. VP]).*
EXAMPLE INPUT	"I love you", *he said.*
SEMANTIC HIERACHY	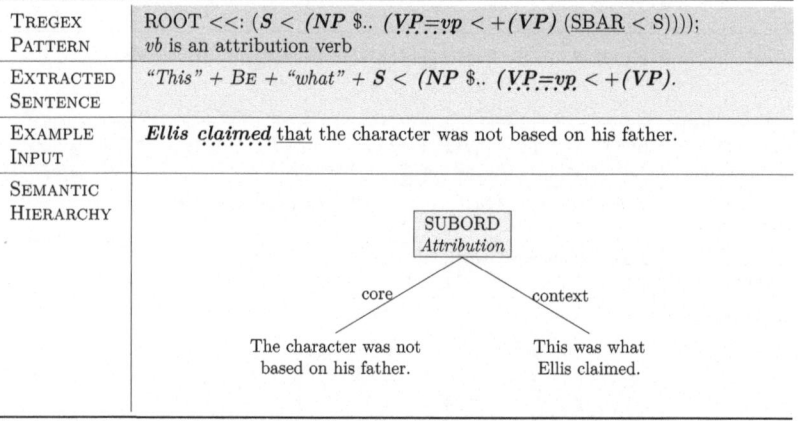

(c) RULE #19: Direct speech with postposed attribution.

TREGEX PATTERN	ROOT <<: *(S < (NP $.. (VP=vp < +(VP) (SBAR < S))));* *vb* is an attribution verb
EXTRACTED SENTENCE	*"This" + BE + "what" + S < (NP $.. (VP=vp < +(VP).*
EXAMPLE INPUT	*Ellis claimed* that the character was not based on his father.
SEMANTIC HIERACHY	

SUBORD
Attribution

core   context

The character was not   This was what
based on his father.   Ellis claimed.

(d) RULE #20: Reported speech with preposed attribution.

this relationship is inverted now. Since we consider the actual statement (e.g., *"He would resign his Senate seat."*) to be more meaningful than the fact that this was pronounced by some entity (e.g., *"This was what Obama announced."*), we tag the subordinate clause containing the uttered proposition as core, while we classify the superordinate attribution assertion as context. Also note that the syntactic pattern of RULE #20 matches the rule that addresses postposed adverbial clauses, i.e. RULE #2. Thus, to ensure that the attribution pattern is favoured in the presence of a verb of reported speech or cognition, RULE #20 is triggered before RULE #2 in the order of execution of the transformation patterns that we specified (see Section 19.1 the online supplemental material).

6.2   Phrasal Disembedding

Clausal disembedding, which is achieved by decomposing coordinate clauses, adverbial clauses, relative clauses and reported speech constructs (see Section 6.1), results in simple sentences, i.e. sentences that each contain exactly one independent clause. However, such sentences may still present a rather complex structure that mixes multiple semantic units. For instance, when simplifying the following input sentence on the clausal level

> *"After graduating from Columbia University in 1983 with a degree in political science, Barack Obama worked as a commmunity organizer in Chicago."*

we end up with two decomposed spans:

- *Barack Obama worked as a community organizer in Chicago.*
- *Barack Obama was graduating from Columbia University in 1983 with a degree in political science.*

Hence, in order to split the input into sentences where each span represents an atomic unit that cannot be further decomposed into meaningful propositions, the TS approach that we propose incorporates phrasal disembedding. According to our goal of generating minimal semantic units, with each of them expressing a complete and indivisible thought, it ensures that phrasal components, too, are transformed into self-contained sentences, in addition to splitting and rephrasing clausal elements, as described above. The resulting semantic hierachy of the example sentence is displayed in Figure 6.3.

```
(1) #1    0    Barack Obama worked as a community organizer.
(1a)              S:TEMPORAL           This was in Chicago.
(1b)              L:TEMPORAL_BEFORE    #2

(2) #2    1    Barack Obama was graduating from Columbia University.
(2a)              S:TEMPORAL           This was in 1983.
(2b)              S:ELABORATION        This was with a degree in political
 science.
```

Figure 6.3 Semantic hierarchy of minimal propositions after clausal and phrasal disembedding

In total, our TS approach addresses seven types of phrasal constructs, including

(1) **coordinate verb phrases,**
(2) **coordinate noun phrase lists,**
(3) **participial phrases,**
(4) **appositive phrases,**
(5) **prepositional phrases,**
(6) **adjectival and adverbial phrases,** as well as
(7) **lead noun phrases.**

In the following, the transformation patterns that were specified for targeting aforementioned linguistic constructs are presented.

6.2.1 Coordinate Verb Phrases

If a verb phrase is made up of multiple coordinate verb phrases, as is the case in the example sentence depicted in Table 6.6, each verb phrase is decomposed and appended to the shared noun phrase to which they refer (Cetto, 2017), thus generating two or more simplified sentences with reduced sentence length. In this way, we aim to increase the minimality of the resulting propositions in the sense that each semantic unit that is expressed in the individual verb phrases ends up in a separate output sentence. Analogous to the rule for splitting and rephrasing coordinate clauses, the textual span that links two coordinate verb phrases is considered as the signal span which is mapped to its corresponding rhetorical relation. Since coordinate verb phrases are of equal status, all of them are labeled as core sentences.

Table 6.6 The transformation rule pattern for disembedding coordinate verb phrases (RULE #21). The pattern in bold represents the part of the input that is extracted from the source and rephrased into a self-contained sentence. The boxed pattern refers to its referent. The underlined pattern (solid line) will be deleted from the remaining part of the input. The pattern that is underlined with a dotted line serves as a cue phrase for the rhetorical relation identification step

TREGEX PATTERN	ROOT <<: (S < (NP $.. (VP < +(VP) (VP > VP ?$.. CC & $.. **VP**))))
EXTRACTED SENTENCE	NP + **VP**.
EXAMPLE INPUT	After Pearlman's bankruptcy, the company emerged unscathed <u>and</u> **was sold to a Canadian company**.
SEMANTIC HIERARCHY	

The transformation rule that defines how to perform this process is specified in terms of a Tregex pattern in Table 6.6. The interested reader may refer to the online supplemental material, where this pattern is illustrated in the form of a phrasal parse tree (see Figure 19.5 in the online supplemental material). Its application on an example sentence is shown in Figure 19.6 in the online supplemental material.

6.2.2 Coordinate Noun Phrase Lists

In accordance with the objective of splitting complex input sentences into a set of minimal semantic units, we compiled a set of rules for breaking up lists of coordinate noun phrases. The specified transformation rules target patterns of coordinate noun phrases within a parent noun phrase. In order to avoid inadvertently mistaking coordinate noun phrases for appositives, we apply a heuristic that is given by the following regular expression:

$$(NP)(, NP)*, ?(and|or)(.+)$$

This pattern is matched only with the topmost noun phrases in subject and object position of the source sentence. By restricting its application to the topmost noun phrases, we compensate for parsing errors that were frequently encountered in deep nested noun phrase structures (Cetto, 2017).

For each identified entity, a simplified sentence is generated according to the rules depicted in Table 6.7. The signal span corresponds to one of the two coordinating conjunctions *"and"* or *"or"* which joins the individual noun phrases, resulting in a "List" relationship. Since the items are linked via a coordinating conjunction, suggesting that they are of equal importance, we tag each split component with a core label. Figure 19.7 in the online supplemental material shows the rule for decomposing coordinate noun phrases in object position in terms of a phrasal parse tree. An example which matches this pattern can be found in Figure 19.8 in the online supplemental material.

6.2.3 Participial Phrases

When two or more actions are carried out simultaneously or immediately one after the other by the same subject, participial phrases are often used to express one of them (e.g., *Knowing that he wouldn't be able to buy food on his journey* "he took large supplies with him.", meaning "As he knew ..."). Note that participial phrases do not contain a subject of their own. Instead, both the subject and the verb of the phrase are replaced by a participle (Martinet and Thomson, 1996). Furthermore, a participial phrase may be introduced by adverbial connectors, such as *"although"*, *"after"* or *"when"* (Abraham, 1985). For an example, see Figure 19.10 in the online supplemental material, which matches the transformation rule specified in Figure 19.9 in the online supplemental material that targets participial phrases in preceding position.

The transformation patterns that were specified for decomposing participial phrases are listed in Table 6.8. For each action expressed in the input, a separate simplified sentence is created. In order to generate an output that is grammatically sound, a paraphrasing stage is required where the participle has to be inflected, if necessary, and linked to the noun phrase that represents the subject it replaces.

Providing some additional piece of information about their respective referent, participial phrases take on a role similar to relative clauses in the semantics of a sentence; the only difference is that they lack a relative pronoun. Therefore, they are sometimes referred to as "reduced relative clauses". Hence, in analogy to relative clauses, the rephrasings resulting from the extracted participial phrases are labelled as context sentences, while the remaining part of the input is tagged

Table 6.7 The transformation rule patterns for splitting coordinate noun phrase lists. A pattern in bold represents the part of the input that is extracted from the source and rephrased into a self-contained sentence. The boxed pattern designates its referent. The underlined part (solid line) will be deleted from the remaining part of the input. The pattern that is underlined with a dotted line serves as a cue phrase for the rhetorical relation identification step

TREGEX PATTERN	ROOT <<: (S < (NP $.. (VP << (NP=np1 < (NP ?$.. CC & $.. NP=np2)))))
EXTRACTED SENTENCE	ROOT <<: (S < (NP $.. (VP << (NP=np2)))).
EXAMPLE INPUT	Demola Aladekomo is a computer engineer, a technology pioneer, an entrepreneur and **a philanthropist.**
SEMANTIC HIERARCHY	

(a) RULE #22: Coordinate noun phrase lists in object position.

TREGEX PATTERN	ROOT <<: (S < (NP=np1 < (NP ?$.. CC & $.. **NP=np2**) $.. VP))
EXTRACTED SENTENCE	ROOT <<: (S < (**NP=np2**) $.. VP).
EXAMPLE INPUT	The intensity of the synthesizer rises before an organ, a bass guitar and **a piano** enter .
SEMANTIC HIERARCHY	

(b) RULE #23: Coordinate noun phrase lists in subject position.

Table 6.8 The transformation rule patterns addressing participial phrases. A pattern in bold represents the part of the input that is extracted from the source and rephrased into a self-contained sentence. An underlined pattern (solid line) will be deleted from the remaining part of the input. The pattern that is underlined with a dotted line serves as a cue phrase for the rhetorical relation identification step. A boxed pattern designates its referent. An italic pattern will be labelled as a context sentence

Tregex Pattern	ROOT <<: (S < VP &<< (NP|PP <, (NP ?$+ PP & $++ (/,/ $+ (*VP [<, (ADVP|PP $+ VBG\|VBN) \| <, VBG\|VBN]* & ?$+ /,/))))
Extracted Sentence	*NP ?$+ PP* + *Be* + *VP [<, (ADVP|PP $+ VBG|VBN) | <, VBG|VBN]*.
Example Input	The Metox , *named after its manufacturer,* *was a high frequency very sensitive radar receiver.*
Semantic Hierarchy	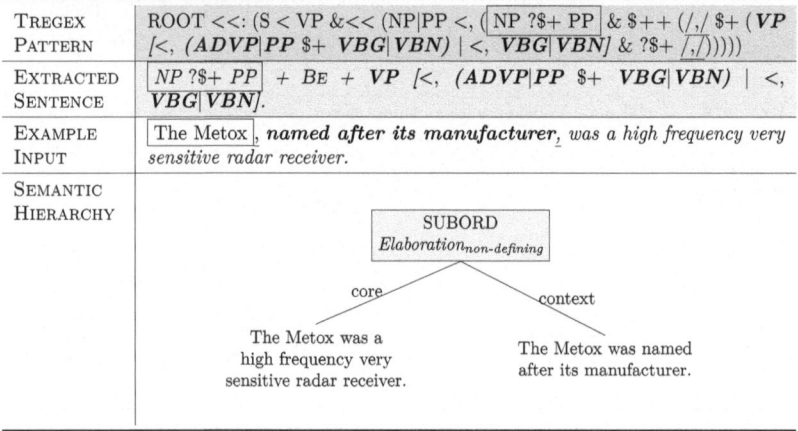

(a) Rule #24: Non-restrictive embedded participial phrases.

(Continued)

as a core sentence, and the split components are connected via an "Elaboration" relation.[6] Analogously to relative clauses, we distinguish between *defining* "Elaboration" relationships for restrictive participial phrases and *non-defining* "Elaboration" relationships for their non-restrictive counterparts.

6.2.4 Appositive Phrases

An appositive is a noun phrase that further characterizes the phrase to which it refers. Similar to relative clauses, appositions can be classified as either restrictive or non-restrictive. Non-restrictive appositives are separate information units, marked by

---

[6] Note that participial phrases introduced by adverbial connectors, such as in the example sentence of Table 6.8d, form an exception of this principle. In such a case, the adverbial connector determines the semantic relationship that holds between the decomposed spans, acting as the cue phrase for identifying the type of rhetorical relation connecting them.

Table 6.8 (Continued)

TREGEX PATTERN	ROOT <<: (S < VP &<< (NP|PP <, ([NP] $+ *(VP [<, (ADVP|PP $+ VBG|VBN) | <, VBG|VBN]*)) & [> (PP ! > S)| > (VP > S)])))
EXTRACTED SENTENCE	[NP] *+ BE + (VP [<, (ADVP|PP $+ VBG|VBN) | <, VBG|VBN]*).
EXAMPLE INPUT	The Muppets at Walt Disney World is [a film] *starring Jim Henson's Muppets at Walt Disney World.*
SEMANTIC HIERARCHY	

(b) RULE #25: Restrictive postposed participial phrases.

TREGEX PATTERN	participialNode = "(___=node [== S=s | == (PP|ADVP < +(PP|ADVP) S=s)]) : (=s <: (VP <<, VBG|VBN))"; ROOT <<: (S < ([NP] $.. (VP < +(VP) (NP|PP $.. " + *participialNode* + "))))
EXTRACTED SENTENCE	[NP] *+ (HAVE) + BE + participialNode.*
EXAMPLE INPUT	[He] served as chief judge from 1987 to 1994, *assuming senior status on November 2, 1995.*
SEMANTIC HIERARCHY	

(c) RULE #26: Non-restrictive postposed participial phrases.

(Continued)

Table 6.8 (Continued)

TREGEX PATTERN	participialNode = "(___=node [== S=s | == (PP|ADVP < +(PP|ADVP) S=s)]) : (=s <: (VP <<, VBG|VBN))";
	ROOT <<: (S < " + *participialNode* + ") : (=node $.. (NP $.. VP))
EXTRACTED SENTENCE	*NP* + *(HAVE)* + BE + *participialNode*.
EXAMPLE INPUT	*Before entering politics,* Donald Trump *was a businessman and a television personality.*
SEMANTIC HIERARCHY	

(d) RULE #27: Non-restrictive preposed participial phrases.

segregation through punctuation (Quirk et al., 1985), such as in *"His main opponent was **Mitt Romney**, the former governor of Massachusetts."*

The pattern for transforming non-restrictive appositions is given in Table 6.9a. It searches for a noun phrase whose immediate right sister is a comma that in turn has another noun phrase as its direct right sibling. In order to avoid inadvertently mistaking coordinate noun phrases for appositives, the following heuristic is applied: from the phrase that is deemed an appositive by matching the pattern described above, we scan ahead, looking one after the other at its sibling nodes in the parse tree. If a conjunction *"and"* or *"or"* is encountered, the analysis of the appositive is rejected (Siddharthan, 2006). In this way, we avoid wrong analyses like: *"Obama has talked about using alcohol, [_appos_ marijuana], and cocaine."*

The second type of appositives, restrictive apposition, does not contain punctuation (Quirk et al., 1985). An example for such a linguistic construct is illustrated in the following sentence: *"Joe Biden was formally nominated by former President **Bill Clinton** as the Democratic Party candidate for vice president."* The pattern in Table 6.9b shows the heuristic that was specified for rephrasing and transforming restrictive appositive phrases into a contextual hierarchy. It defines a regular

Table 6.9 The transformation rule patterns addressing appositive phrases. A pattern in bold represents the part of the input that is extracted from the source and rephrased into a self-contained sentence. An underlined pattern (solid line) will be deleted from the remaining part of the input. A boxed pattern designates its referent. An italic pattern will be labelled as a context sentence

TREGEX PATTERN	ROOT <<: (S < VP & << (NP $+ (/,/ $+ (*NP* !$ CC & ?$+ /,/))))
EXTRACTED SENTENCE	NP + BE + *NP*.
EXAMPLE INPUT	The president of Lithuania , *Antanas Smetona*, proposes armed resistance.
SEMANTIC HIERARCHY	

(a) RULE #28: Non-restrictive appositive phrases.

TREGEX PATTERN	*(((PRP\\$|DT)\\ s)*(JJ\\ s)*((NN|NNS|NNP|NNPS)\\ s))+(((CC|IN)\\ s)((PRP\\$|DT)\\ s)*(JJ\\ s)*((NN|NNS|NNP|NNPS)\\ s))** followed by a named entity
EXTRACTED SENTENCE	*named entity* + BE + *regex*.
EXAMPLE INPUT	The regional government was moved from *the old Cossack capital* Novocherkassk to Rostov.
SEMANTIC HIERARCHY	

(b) RULE #29: Restrictive appositive phrases.

Table 6.10 The transformation rule patterns addressing prepositional phrases. A pattern in bold represents the part of the input that is extracted from the source and rephrased into a self-contained sentence. An underlined pattern (solid line) will be deleted from the remaining part of the input. An italic pattern will be labelled as a context sentence

TREGEX PATTERN	ROOT <<: (S < +(S|VP) (VP < (***PP*** $– NP|PP)) & < VP)
EXTRACTED SENTENCE	*"This"* + BE + PP .
EXAMPLE INPUT	Brick enabled the construction of permanent buildings *in regions of India where the harsher climate precluded the use of mud bricks.*
SEMANTIC HIERARCHY	

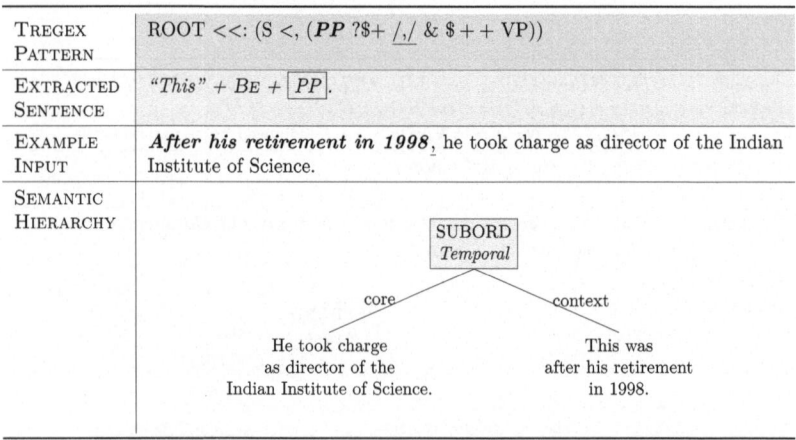

(a) RULE #30: Prepositional phrases that act as complements of verb phrases.

TREGEX PATTERN	ROOT <<: (S <, (***PP*** ?$+ /,/ & $ + + VP))
EXTRACTED SENTENCE	*"This"* + BE + PP .
EXAMPLE INPUT	***After his retirement in 1998****,* he took charge as director of the Indian Institute of Science.
SEMANTIC HIERARCHY	

SUBORD
Temporal

core   context

He took charge
as director of the
Indian Institute of Science.

This was
after his retirement
in 1998.

(b) RULE #31: Preposed prepositional phrases offset by commas.

(Continued)

Table 6.10 (Continued)

TREGEX PATTERN	ROOT <<: (S < VP & << (/,/ $+ (**PP** ?$+ /,/)))
EXTRACTED SENTENCE	"This" + BE + PP .
EXAMPLE INPUT	It later became a Roman town in the province of Africa, *before its eventual abandonment around 9th to 10th century*.
SEMANTIC HIERARCHY	

SUBORD
Temporal

core   context

It later became   This was before
a Roman town in   its eventual abandonment
the province of Africa. around 9th to 10th century.

(c) RULE #32: Postposed and embedded prepositional phrases offset by commas.

Table 6.11 The transformation rule patterns addressing adjectival and adverbial phrases. The pattern in bold represents the part of the input that is extracted from the source and rephrased into a self-contained sentence. The underlined pattern (solid line) will be deleted from the remaining part of the input. The italic pattern will be labelled as a context sentence

TREGEX PATTERN	ROOT <<: (S <, (**ADJP|ADVP** $+ (/,/ $++ VP)))
EXTRACTED SENTENCE	"This" + BE + **ADJP|ADVP**.
EXAMPLE INPUT	*Meanwhile*, unemployment in France threw skilled workers down to the level of the proletariat.
SEMANTIC HIERARCHY	

SUBORD
Elaboration

core   context

Unemployment in France
threw skilled workers This was
down to the level meanwhile.
of the proletariat.

(a) RULE #33: Preposed adjectival and adverbial phrases offset by commas.

(Continued)

Table 6.11 (Continued)

Tregex Pattern	ROOT <<: (S < VP & << (/,/ \$+ (***ADJP***|***ADVP*** ?\$+ /,/)))
Extracted Sentence	*"This"* + Be + ***ADJP***|***ADVP***.
Example Input	Gustafsson lived a normal life until 2004, *almost 44 years after the event.*
Semantic Hierarchy	

(b) Rule #34: Postposed and embedded adjectival and adverbial phrases offset by commas.

expression that searches for a noun or a proper noun (or a coordinate sequence thereof), optionally with a combination of prepending adjectives, determiners, and possessive pronouns. This string must be followed by a named entity expression.

According to the rules listed in Table 6.9, the appositive phrases are extracted from the input and transformed into stand-alone sentences. Representing parenthetical information, they are labelled as context sentences, while the reduced source sentences receive a core tag. Since appositions commonly provide additional information about the entity to which they refer, we link the split sentences by an "Elaboration" relation.

6.2.5 Prepositional Phrases

A prepositional phrase is composed of a preposition and a complement in the form of a noun phrase (e.g., *"on the table"*, *"in terms of money"*), a nominal *wh*-clause (e.g., *"from what he said"*) or a nominal *-ing* clause (e.g., *"by signing a peace treaty"*). They may function as a postmodifier in a noun phrase or an adverbial phrase, or act as a complement of a verb phrase or an adjective (Quirk et al., 1985).

We defined a set of rules for decomposing two variants of prepositional phrases (see Table 6.10). First, we address prepositional phrases that are offset by commas, e.g. *"At the 2004 Democratic National Convention, Obama delivered the keynote address."* or *"When they moved to Washington, D.C., in January 2009, the girls started at the Sidwell Friends School."* (see the patterns in Table 6.10b and 6.10c) Such linguistic constructs typically provide some piece of background information that may be extracted from the source sentence without corrupting the meaning of the input. Second, we specified a pattern for transforming prepositional phrases that act as complements of verb phrases, such as in *"Obama formally **announced** his candidacy in January 2003."* or *"Obama **defeated** John McCain in the general election."* (see the pattern in Table 6.10a) Often, such verb phrase modifiers represent optional constituents that contribute no more than some form of additional information which can be eliminated, resulting in a simplified source sentence that is still both meaningful and grammatically sound.

However, automatically distinguishing optional verb phrase modifiers from those that are required in order to form a syntactically and semantically well-formed sentence is challenging. For instance, consider the following input: *"Radio France is **headquartered** in Paris' 16th arrondissement."* When separating out the prepositional phrase that modifies the verb phrase, the output (*"Radio France is headquartered."*) is overly terse. To avoid this, we implemented the following heuristic: prepositional phrases that complement verb phrases are decomposed only if there is another constituent left in the object position of the verb phrase. In that way, we ensure that the resulting output presents a regular subject-predicate-object structure that is still meaningful and at the same time easy to process for downstream Open IE applications.

Representing some piece of background information, the extracted prepositional phrases are labelled as context sentences, while the remaining part of the input is tagged as a core sentence. Since prepositional phrases commonly express either temporal or spatial information, the rhetorical relation identification step is carried out based on named entities. For this purpose, we iterate over all the tokens contained in the extracted phrase. If we encounter a named entity of the type "LOCATION", we join the decomposed sentences by a "Spatial" relationship. In case of a named entity of the class "DATE", the split sentences are linked through a "Temporal" relation.

6.2.6 Adjectival and Adverbial Phrases

An adjectival phrase is a phrase whose head is an adjective that is optionally complemented by a number of dependent elements. It further characterizes the noun phrase it is modifying. Similarly, an adverbial phrase consists of an adverb as its head, together with an optional pre- or postmodifying complement (Brinton, 2000; Quirk et al., 1985). The rules for detecting and transforming these types of linguistic constructs are listed in Table 6.11 in terms of Tregex patterns.

Note that our TS approach is limited to extracting adjectival and adverbial phrases that are offset by commas. Sentences that contain attributive adjectives or adverbial modifiers that are not separated through punctuation from the main clause (e.g., *"He has a red car."*, *"I usually go to the lake."*) are not simplified, as the underlying sentence structure typically already presents a regular subject-predicate-object order. Therefore, the sentence is already easy to process, outweighing the risk of introducing errors when attempting to further simplify the input by transforming the adjective or adverb into a self-contained sentence.

Whenever one of the patterns from Table 6.11 matches a sentence's phrasal parse tree, the adjectival or adverbial phrase, respectively, is extracted from the input and turned into a stand-alone simplified sentence by prepending the canoncial phrase *"This is/was"*. As it commonly expresses a piece of background information, it is labelled as a context sentence. Containing the key information of the input, the remaining part of the source receives a core tag. Aside from setting up a contextual hierarchy between the split elements, a semantic relationship between them is established. For this purpose, the decomposed sentences are connected with an "Elaboration" relation, which is selected from the classes of rhetorical relations due to the fact that adjectival and adverbial phrases usually provide additional details about the event described in the respective main clause.

6.2.7 Lead Noun Phrases

Occasionally, sentences may start with an inserted noun phrase, which generally indicates a temporal expression. Hence, such a phrase usually represents background information that can be eliminated from the main sentence without resulting in a lack of key information. This is achieved by applying the transformation rule displayed in Figure 19.11 in the online supplemental material. The corresponding Tregex pattern is specified in Table 6.12. Since the information expressed in the leading noun phrase commonly represents a piece of minor background information, the resulting extraction is labelled as a context sentence, while the remaining part from the input is tagged as a core sentence. The two simplified components are then linked through a "Temporal" relation. In that way, a semantic hierarchy is established between the decomposed sentences. Figure 19.12 in the online supplemental material illustrates an example.

Table 6.12 The transformation rule pattern for splitting and rephrasing leading noun phrases (RULE #35). The pattern in bold represents the part of the input that is extracted from the source and turned into a self-contained sentence. The underlined pattern (solid line) will be deleted from the remaining part of the input. The italic pattern will be labelled as a context sentence

TREGEX PATTERN	ROOT <<: (S <, (***NP-TMP/NP*** \$+ (/,/ \$+ NP & \$ + + VP)))
EXTRACTED SENTENCE	*"This"* + BE + ***NP-TMP/NP***.
EXAMPLE INPUT	***Six days later***, NATO took over leadership of the effort.
SEMANTIC HIERARCHY	 SUBORD *Temporal* core context NATO took over This was six leadership of the effort. days later.

Transformation Process   7

In the following, we will first present a formal representation of the semantic hierarchy of minimal propositions that is generated by our proposed discourse-aware TS
approach. Next, we will integrate the different components of our framework that
were presented in the previous sections by describing the transformation algorithm
we specified. Finally, we will demonstrate it in use, detailing the transformation
process on an example sentence.

7.1   Data Model: Linked Proposition Tree

The algorithm we propose takes a complex sentence as input and recursively transforms it into a semantic hierarchy of minimal propositions. The output is represented
as a linked proposition tree. Its basic structure is depicted in Figure 7.1. A linked
proposition tree is a labeled binary tree[1] $LPT = (V, E)$.

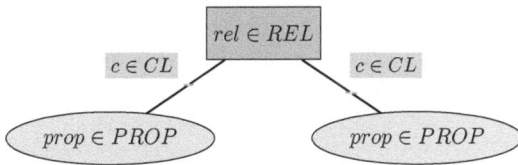

Figure 7.1 Basic structure of a linked proposition tree LPT. It represents the data model of
the semantic hierarchy of minimal propositions

---

[1] In rare cases, a node may have more than two children, e.g. when the input sentence contains
a list of noun phrases enumerating more than two entities.

© The Author(s), under exclusive license to Springer Fachmedien Wiesbaden GmbH,   143
part of Springer Nature 2022
C. Niklaus, *From Complex Sentences to a Formal Semantic Representation using
Syntactic Text Simplification and Open Information Extraction*,
https://doi.org/10.1007/978-3-658-38697-9_7

Let $V \in \{REL, PROP\}$ be the set of nodes, where $PROP$ is the set of leaf nodes denoting the set of minimal propositions.

A $prop \in PROP$ is a triple $(s, v, o) \in CT$, where CT represents the set of clause types: $CT = \{SV, SVA, SVC, SVO, SVOO, SVOA, SVOC\}$ (see Table 4.1).

$s \in S$ denotes a subject, $v \in V$ a verb and $o \in \{O, A, C, OO, OA, OA, \emptyset\}$ a direct or indirect object (O), adverbial (A) or complement (C) (or a combination thereof).

Hence, a minimal proposition $prop \in PROP$ is a simple sentence that is reduced to its clause type.[2] Thus, it represents a minimal unit of coherent information where all optional constituents are discarded, resulting in an utterance that expresses a single event consisting of a predicate and its core arguments (see Section 4.1). Accordingly, a minimal proposition presents a simple and regular structure that is easier to process and analyze for downstream Open IE tasks.

Furthermore, let $REL = \{Contrast, List, Disjunction, Cause, Result, Background, Condition, Elaboration, Attribution, Purpose, Temporal-Before, Temporal-After, Temporal, Spatial, Unknown-Coordination, Unknown-Subordination\}$ be the set of rhetorical relations, comprising the set of inner nodes.

A $rel \in REL$ represents the semantic relationship that holds between its child nodes. It reflects the semantic context of the associated propositions $prop \in PROP$. In that way, the coherence structure of the input is preserved.

Finally, let $E \in CL$ be the set of constituency labels, with $CL \in \{core, context\}$.

A $c \in CL$ represents a labeled edge that connects two nodes $V \in LPT$. It enables the distinction between core information and less relevant contextual information. In that way, hierarchical structures between the split propositions $prop \in PROP$ are captured.

---

[2] Beyond that, a specified set of phrasal units is also extracted in order to avoid overly complex subclausal elements (see Sections 4.1.1 and 6.2). For the sake of simplicity, this is not reflected in the above notation.

7.2 Transformation Algorithm

After having explained in detail both the individual subtasks that are carried out in our context-preserving TS approach (see Chapters 4 and 5) and the full set of hand-crafted transformation patterns (see Chapter 6), we will now describe the transformation algorithm (see Algorithm 1) that brings everything together. It takes a sentence as input and applies the set of linguistically principled transformation rules to recursively convert it into a semantic hierarchy of minimal propositions. In doing so, it

(1) splits up and rephrases the input into a set of *minimal propositions*,
(2) sets up a *contextual hierarchy* between the split components (a), and identifies the *semantic relationship* that holds between them (b).

In that way, a linked proposition tree is created in terms of a set of hierarchically ordered and semantically interconnected sentences that present a simplified syntax.

Initialization In the initialization step (1–8), the linked proposition tree *LPT* is instantiated with the source sentence. It is represented as a single leaf node that has an unlabeled edge to the root node. For an example, see Figure 7.3.

Tree Traversal Next, the linked proposition tree *LPT* is recursively traversed, splitting up the input in a top-down approach (9–34). Starting from the root node, the leaves are processed in depth-first order. For every leaf (11), we check if its phrasal parse tree matches one of the transformation patterns (13). The rules are applied in a fixed order that was empirically determined (see Section 4.3). The first pattern that matches the proposition's parse tree is executed (14).

Algorithm 1 Transform into Semantic Hierarchy of Minimal Propositions

---

Require: complex source sentence *str*
Ensure: linked proposition tree *tree*

1: **function** INITIALIZE(*str*)
2:    *new_leaves* ← source sentence *str*
3:    *new_node* ← create a new parent node for *new_leaves*
4:    *new_node.labels* ← None
5:    *new_node.rel* ← ROOT
6:    linked proposition tree *tree* ← initialize with *new_node*
7:    **return** *tree*
8: **end function**

9: **procedure** TRAVERSETREE(*tree*)
10:    ▷ Process leaves (i.e. propositions) from left to right
11:    **for** *leaf* in *tree.leaves* **do**
12:       ▷ Check transformation rules in fixed order
13:       **for** *rule* in *TRANSFORM_RULES* **do**
14:          **if** *match* **then**
15:             ▷ **(a) Sentence Splitting**
16:             *simplified_propositions* ← decompose *leaf* into a set of simplified propositions
17:             *new_leaves* ← convert *simplified_propositions* into leaf nodes
18:             ▷ **(b) Constituency Type Classification**
19:             *new_node* ← create a new parent node for *new_leaves*
20:             *new_node.labels* ← link each leaf in *new_leaves* to *new_node* and label each edge
21:                with the leaf's constituency type $c \in CL$
22:             ▷ **(c) Rhetorical Relation Identification**
23:             *cue_phrase* ← extract cue phrase from *leaf.parse_tree*
24:             *new_node.rel* $\in REL$ ← match *cue_phrase* against a predefined set of rhetorical
25:                cue words
26:             ▷ Update Tree
27:             *tree.replace(leaf, new_node)*
28:             ▷ Recursion
29:             TRAVERSETREE(*tree*)
30:          **end if**
31:       **end for**
32:    **end for**
33:    **return** *tree*
34: **end procedure**

(a) Sentence Splitting In a first step, the current proposition is decomposed into a set of shorter utterances that present a more regular structure (16). This is achieved through disembedding clausal or phrasal components and converting them into stand-alone sentences. Accordingly, the transformation rule encodes both the splitting point(s) and the rephrasing procedure for reconstructing grammatically sound sentences (see Chapter 4). Each split will result in two[3] sentences with a simpler syntax. They are represented as leaf nodes in the linked proposition tree (17) (see subtask (a) in Figure 7.2). To establish a semantic hierarchy between the split spans, two further subtasks are carried out.

(b) Constituency Type Classification To set up a contextual hierarchy between the split sentences, the transformation rule determines the constituency type $c \in CL$ of the leaf nodes that were created in the previous step (19–21). To differentiate between *core* sentences that contain the key message of the input and *contextual* sentences that provide additional information about it, the transformation pattern encodes a simple syntax-based method. Based on the assumption that subordinations commonly express background information, simplified propositions resulting from subordinate clausal or phrasal elements are classified as context sentences, while those emerging from their superordinate counterparts are labelled as core sentences. Coordinations, too, are flagged as core sentences, since they are of equal status and typically depict the main information of the input (see subtask (b) in Figure 7.2). For details, see Section 5.1.

(c) Rhetorical Relation Identification To preserve the semantic relationship between the simplified propositions, we classify the rhetorical relation $rel \in REL$ that holds between them. For this purpose, we utilize a predefined list of rhetorical cue words (Taboada and Das, 2013). To infer the type of rhetorical relation, the transformation pattern first extracts the cue phrase of the given sentence (23). It is then used as a lexical feature for classifying the semantic relationship that connects the split propositions (24, 25). For example, the rule in Table 6.2a specifies that the phrase *"although"* is the cue word in the source sentence of Table 1.1, which is mapped to a "Contrast" relationship according to the findings in Taboada and Das (2013) (see subtask (c) in Figure 7.2). More details can be found in Section 5.2.

---

[3] When joining more than two clauses or phrases, respectively, the rules targeting coordinations (i.e., coordinate clauses, coordinate verb phrases and coordinate noun phrases) result in more than two simplified sentences. However, this is comparatively rarely the case.

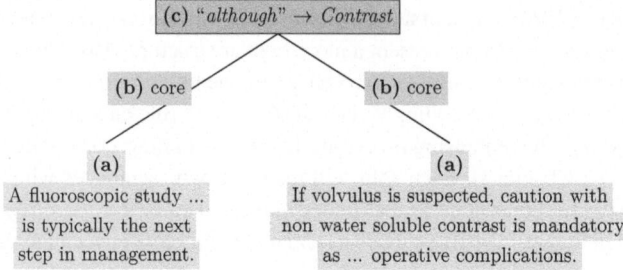

Figure 7.2 Semantic hierarchy after the first transformation pass. (**Subtask a**) The source sentence is split up and rephrased into a set of syntactically simplified sentences. (**Subtask b**) Then, the split sentences are connected with information about their constituency type to establish a contextual hierarchy between them. (**Subtask c**) Finally, by identifying and classifying the rhetorical relation that holds between the simplified sentences, their semantic relationship is captured

Recursion Next, the linked proposition tree LPT is updated by replacing the leaf node that was processed in this run with the newly generated subtree (27). It is composed of the simplified propositions *(leaf nodes)*, their semantic relationship $rel \in REL$ *(inner node)* and constituency labels $c \in CL$ *(edges)*. Figure 7.2 depicts the result of the first transformation pass on the example sentence from Table 1.1. The resulting leaf nodes are then recursively simplified in a top-down fashion (29).

Termination The algorithm terminates when no more rule matches the set of simplified propositions $prop \in PROP$ in the leaf nodes. It outputs the source sentence's linked proposition tree LPT (33), representing its semantic hierarchy of minimal semantic units. In that way, the input is transformed into a set of hierarchically ordered and semantically interconnected sentences that present a simplified syntax. Figure 7.14 shows the final linked proposition tree LPT of our example sentence.

7.3   Transformation Example

In the following, a step-by-step example is given, illustrating the transformation process on the complex sentence from Table 1.1. With the help of the 35 hand-crafted grammar rules detailed in Chapter 6, the input is recursively transformed into a linked proposition tree, i.e. a set of hierarchically ordered and semantically interconnected sentences in the form of minimal propositions that present a simple and regular structure.

Initialization In the initialization step, the linked proposition tree LPT is instantiated. At this point, it consists of the source sentence, which is represented as a single leaf node, and an unlabeled edge to the root node (see Figure 7.3).

ROOT

A fluoroscopic study which is known as an upper gastrointestinal series is typically the next step in management, although if volvulus is suspected, caution with non water soluble contrast is mandatory as the usage of barium can impede surgical revision and lead to increased post operative complications.

(0)

Figure 7.3 Initialization of the linked proposition tree LPT

Rule application #1 The execution phase of the transformation algorithm starts by examining the phrasal parse tree of the input sentence (see Figure 7.4).

The algorithm passes through the list of specified transformation rules in a fixed order (see Section 4.3), comparing the patterns against the sentence's parse structure. In our example, the first rule that matches is RULE #2, which targets sentences that contain a postposed adverbial clause. It simplifies the input by disembedding the subordinate clause S' and turning it into a stand-alone sentence. The pattern defines the constituent that precedes S' within the subordinate clause $SBAR$ as the signal span for the rhetorical relation identification step. Hence, in our case, the cue word is *"although"*, which is mapped to a "Contrast" relationship. Normally, subordinate clauses would be classified as a context sentence, in accordance with their syntactic constituency. However, in the theory of RST, text spans that present a "Contrast" relationship are considered both as a nucleus. Therefore, both decomposed spans are labeled as core sentences in our example, constituting a new coordination node in the linked proposition tree. The resulting simplified sentences, represented as leaf nodes, are:

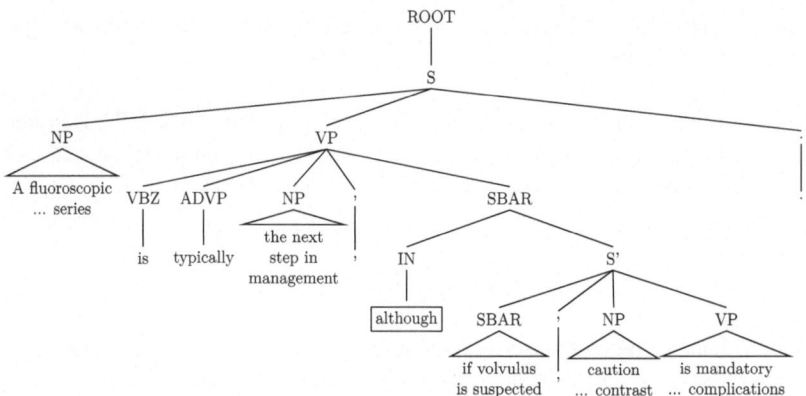

Figure 7.4 Phrasal parse tree of the input sentence

#1 Core: *A fluoroscopic study which is known as an upper gastrointestinal series is typically the next step in management.*
#2 Core: *If volvulus is suspected, caution with non water soluble contrast is mandatory as the usage of barium can impede surgical revision and lead to increased post operative complications.*

The corresponding semantic hierarchy after the first transformation pass is depicted in Figure 7.5.

Rule application #2 The algorithm continues with processing the split sentences, starting with sentence #1. The first rule that matches its phrasal parse tree (see Figure 7.6) is RULE #15, which aims at extracting restrictive relative clauses. It is triggered by the relative pronoun *"which"* in the *wh*-noun phrase $WHNP$.

During the rule application, the relative clause S' is disembedded from the input. In order to generate a grammatically sound output sentence, it is attached to its refer-ring phrase NP'. Since the relative clause provides identifying information about its antecedent, an "Elaboration" relationship is established between the split com-ponents, which is defined more precisely by adding the adjunct "defining". In that way, we clarify that the extracted clause is an integral part of the phrasal component it modifies. The constituency labels are derived from the syntactic constituency of the source sentence, creating a new subordination node with S' as the contextual sentence and the remaining part of the input as the associated core sentence:

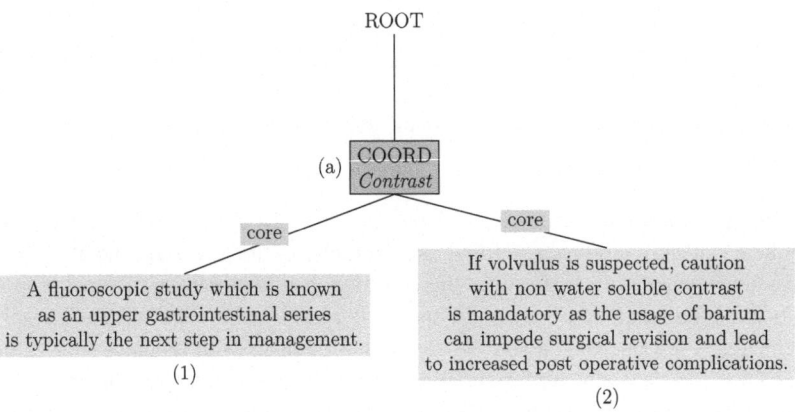

Figure 7.5 Semantic hierarchy after the first transformation pass

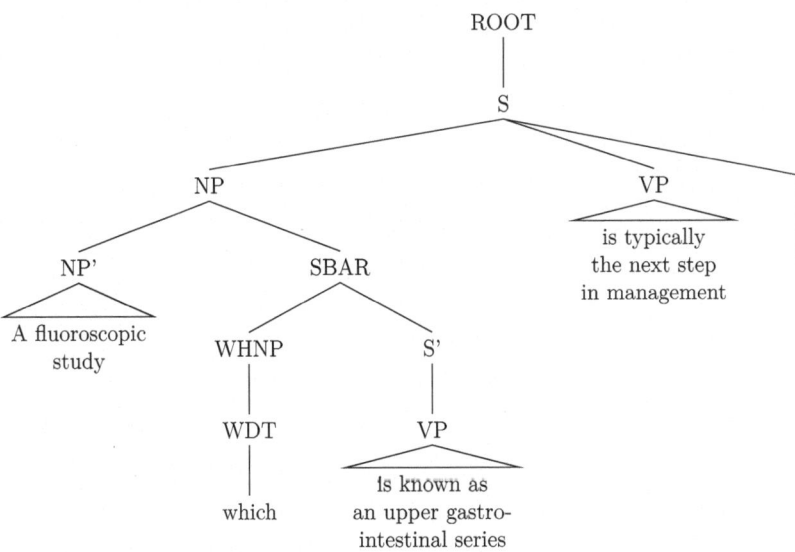

Figure 7.6 Phrasal parse tree of the split sentence #1

#1.1   Core: *A fluoroscopic study is typically the next step in management.*
#1.2   Context: *This fluoroscopic study is known as an upper gastrointestinal series.*

The resulting semantic hierarchy after the second transformation pass is depicted in Figure 7.7.

Rule application #3 Following a depth-first strategy, the newly generated simplified sentences are processed first. Accordingly, in the next step, the algorithm examines sentence #1.1. However, this sentence already presents a minimal proposition that does not match any of the specified transformation rules. Therefore, this sentence is left unchanged and the algorithm continues with sentence #1.2. This sentence, too, cannot be further simplified according to our transformation patterns. Hence, the algorithm proceeds with examining the phrasal parse tree of sentence #2 (see Figure 7.8). Here, the rule for splitting sentences containing a preposed adverbial clause (RULE #3) is the first to match.

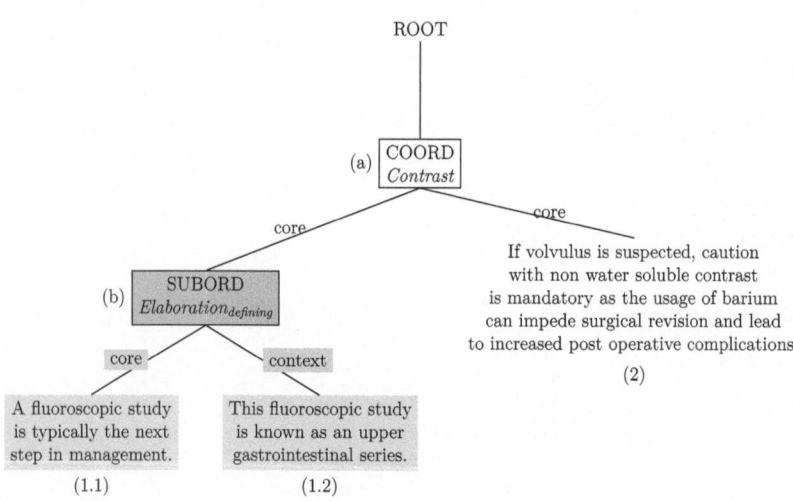

Figure 7.7 Semantic hierarchy after the second transformation pass

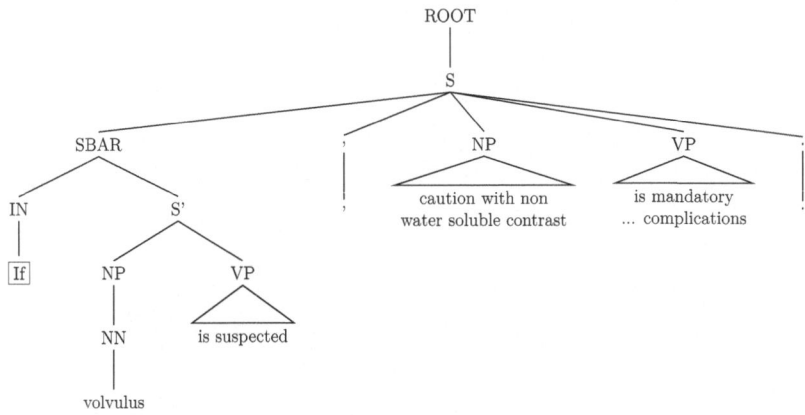

Figure 7.8 Phrasal parse tree of the split sentence #2

Analogous to the first rule application in our example, the input is simplified by decomposing the subordinate clause S' and converting it into a self-contained sentence. The element preceding S' within the subordinate clause $SBAR$, *"if"*, represents the signal span for identifying the type of relationship that connects the split spans. It is classified as a "Condition" relation. Representing some piece of background information, the extracted adverbial clause S' is labeled as a context sentence, while the remaining part of the input receives a core tag. In that way, a new subordination node is created, linking the simplified sentences via a "Condition" relationship:

#2.1 Context: *Volvulus is suspected.*

#2.2 Core: *Caution with non water soluble contrast is mandatory as the usage of barium can impede surgical revision and lead to increased post operative complications.*

The corresponding semantic hierarchy after the third transformation pass is depicted in Figure 7.9.

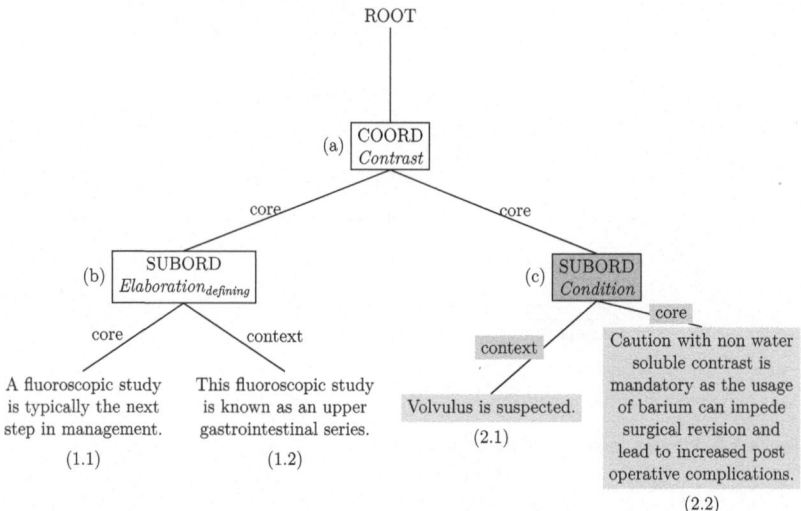

Figure 7.9 Semantic hierarchy after the third transformation pass.

Rule application #4 Sentence #2.1 presents a simple structure that cannot be further simplified by one of our specified rule patterns. Thus, the algorithm proceeds with sentence #2.2. The first rule that matches its phrasal parse tree (see Figure 7.10) is RULE #2, which simplifies sentences by disembedding postposed adverbial clauses.

It extracts the subordinate clause S' from the source sentence, transforming it into a simplified stand-alone sentence. In accordance with its syntactic constituency, it is labeled as a context sentence, whereas the remaining textual span of the input is classified as a core sentence, thus creating a new subordination node in the linked proposition tree. Here, the signal span for the rhetorical relation identification step is *"as"*, which is mapped to a "Background" relationship. Hence, the resulting simplified sentences are:

#2.2.1 Core: *Caution with non water soluble contrast is mandatory.*
#2.2.2 Context: *The usage of barium can impede surgical revision and lead to increased post operative complications.*

The semantic hierarchy resulting from the fourth transformation pass is depicted in Figure 7.11.

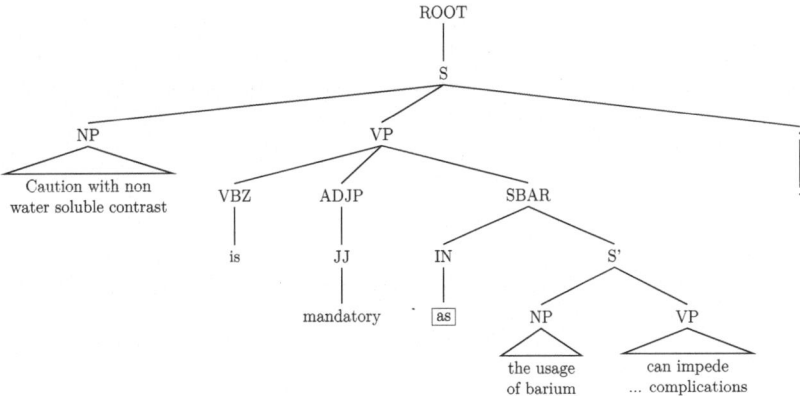

Figure 7.10 Phrasal parse tree of the split sentence #2.2

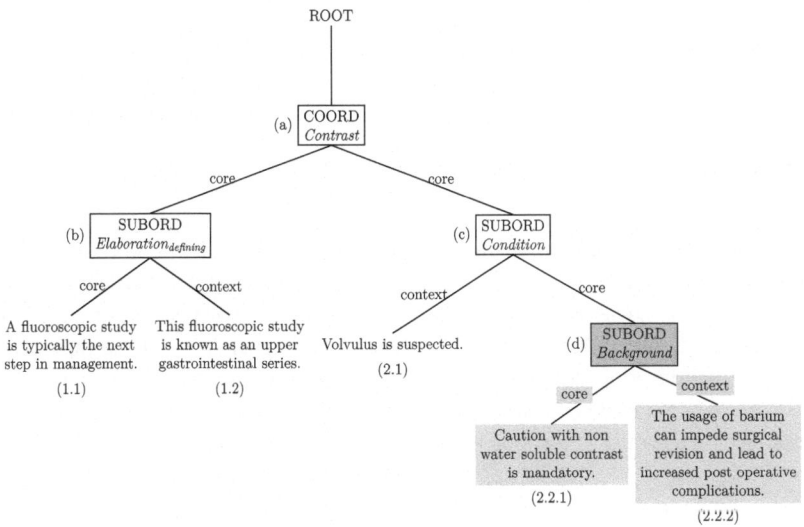

Figure 7.11 Semantic hierarchy after the fourth transformation pass

Rule application #5 Since sentence #2.2.1 does not match any of the specified transformation rules, the algorithm does not further decompose it. Instead, sentence #2.2.2 is processed. The first rule that matches its phrasal parse tree (see Figure 7.12) is RULE #21, which splits coordinate verb phrases.

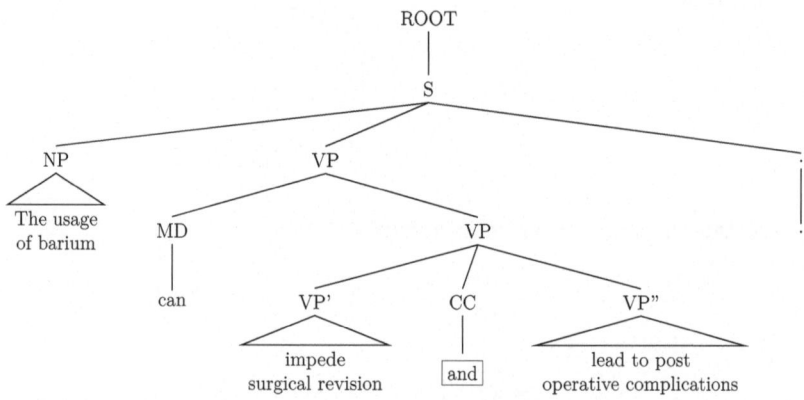

Figure 7.12 Phrasal parse tree of the split sentence #2.2.2.

It is triggered by the coordinating conjunction CC that links the two verb phrases VP' and VP''. Each verb phrase is disembedded and attached to the shared noun phrase NP, thus generating two sentences with a simplified syntax. The conjunction connecting the verb phrases, *"and"*, is used as the signal span for the rhetorical relation identification, which is mapped to a "List" relationship. Since coordinate verb phrases are of equal status, both of them are labeled as core sentences, thus establishing a coordination node in the linked proposition tree.

The split sentences in this step of the rule application are:

#2.2.2.1 Core: *The usage of barium can impede surgical revision.*
#2.2.2.2 Core: *The usage of barium can lead to increased post operative complications.*

The corresponding semantic hierarchy after the fifth transformation pass is depicted in Figure 7.13.

Termination When no more transformation pattern matches the phrasal parse tree of the simplified sentences, the algorithm stops and outputs the generated linked proposition tree LPT, representing the semantic hierarchy of minimal propositions. The final tree of our running example is shown in Figure 7.14.[4] Its leaf nodes represent the minimal propositions $prop \in PROP$ that were generated during the transformation process. Internal nodes are represented by boxes that contain the constituency type (in uppercase letters), as well as the identified rhetorical relation $rel \in REL$ (in italic) that links the associated split sentences. An internal node with

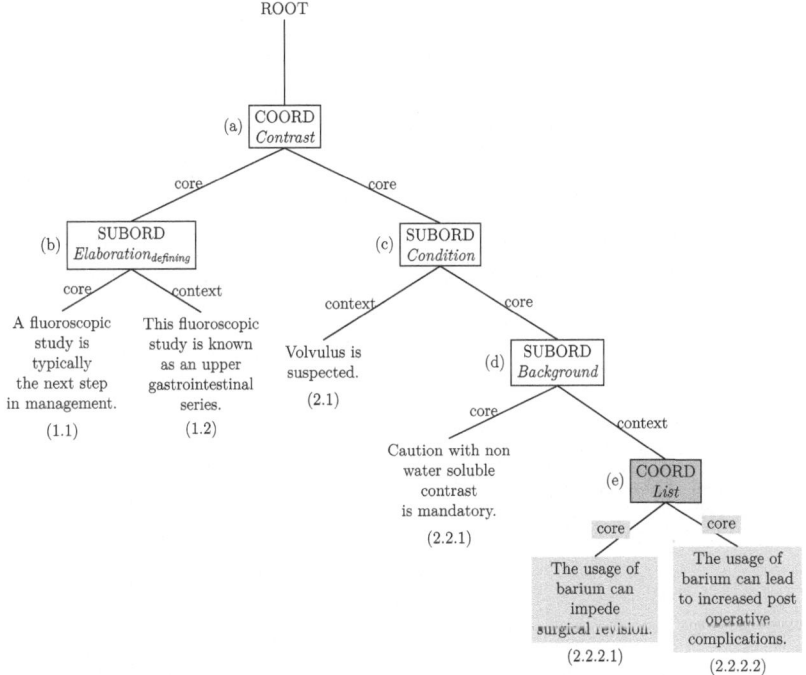

Figure 7.13 Semantic hierarchy after the fifth transformation pass.

---

[4] The resulting minimal propositions each match one of the clause types specified in Section 4.1.1: SV (2.1), SVA (1.2, 2.2.2.1 and 2.2.2.2), SVC (2.2.1) and SVO (1.1).

a coordinate constituency type is called a coordination node in our approach, while an internal node with a subordinate constituency type is referred to as a subordination node. Every node of the tree is connected to its parent by an edge that presents either a core or a context label $c \in CL$. Subordination nodes always have exactly two children that are connected by a core edge (representing the superordinate span) and a context edge (representing the subordinate span). Coordination nodes, too, typically have two child nodes. However, on rare occasions, more than two children are possible, e.g. when the input contains a list of noun phrases enumerating more than two entities.

Two-layered Representation To facilitate its processing in downstream Open IE applications, the linked proposition tree LPT is converted into a two-layered light-

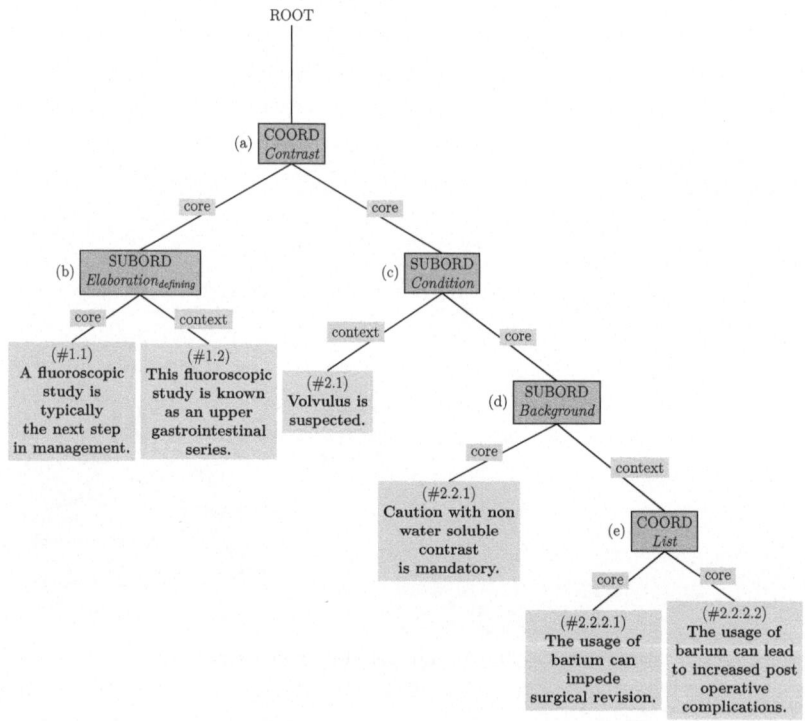

Figure 7.14 Final linked proposition tree LPT of the example sentence

weight semantic representation in the form of core sentences and accompanying contexts that is both machine processable (see Figure 12.5) and human readable (see Figure 12.3). More details on how the tree structure from Figure 7.14 is transformed into a two-layered representation of simplified sentences can be found in Section 12.1.[5]

---

[5] Note that in Section 12.1, we explicitly address the output generated in the relation extraction stage, resulting in a set of tuples (id, cl, t, C_S, C_L), where t represents a relational tuple (arg_1, rel, arg_2) extracted from a simplified sentence s. Thus, when not carrying out this last step and instead terminating after the simplification process, the output is represented in the form of a tuple (id, cl, s, C_S, C_L) for each simplified sentence s.

Sentence Splitting Corpus

8

We leverage our proposed approach for structural TS to compile a new sentence splitting corpus, MINWIKISPLIT,[1] that is composed of 203K pairs of aligned complex source and simplified target sentences. We improve over existing sentence splitting corpora (see Section 2.1.2.2) by gathering a large-scale, syntactically diverse dataset where each complex source sentence is broken down into a set of minimal propositions, i.e. a sequence of sound, self-contained utterances with each of them presenting a minimal semantic unit that cannot be further decomposed into meaningful propositions. This corpus may be useful for developing data-driven sentence splitting approaches that learn how to transform sentences with a complex linguistic structure into a fine-grained representation of short sentences that present a simple and more regular syntax. This output may then serve as an intermediate representation which is easier to process for downstream applications (Narayan et al., 2017; Bouayad-Agha et al., 2009; Štajner and Popović, 2018) and may thus facilitate and improve their performance.

Corpus Construction MINWIKISPLIT is a large-scale sentence splitting corpus consisting of 203K complex source sentences and their simplified counterparts in the form of a sequence of minimal propositions. It was created by running our reference syntactic TS implementation DISSIM over the one million complex input sentences from the WIKISPLIT corpus. As detailed in Chapter 6, DISSIM applies a small set of 35 hand-written transformation rules to decompose a wide range of linguistic constructs, including both clausal components (coordinations, adverbial clauses, relative clauses and reported speech) and phrasal elements (appositions,

---

[1] The MINWIKISPLIT dataset is publicly released under https://github.com/Lambda-3/MinWikiSplit.

C. Niklaus, *From Complex Sentences to a Formal Semantic Representation using Syntactic Text Simplification and Open Information Extraction*, https://doi.org/10.1007/978-3-658-38697-9_8

Table 8.1 Example instances from MINWIKISPLIT. A complex source sentence is broken down into a sequence of syntactically simplified sentences by decomposing clausal and phrasal elements and transforming them into self-contained utterances, resulting in a set of minimal propositions

Complex source	The house was once part of a plantation and it was the home of Josiah Henson, a slave who escaped to Canada in 1830 and wrote the story of his life.
MINWIKISPLIT	(1) **The house was once part of a plantation.**
	(2) **It was the home of Josiah Henson.**
	(3) **Josiah Henson was a slave.**
	(4) **This slave escaped to Canada.**
	(5) **This was in 1830.**
	(6) **This slave wrote the story of his life.**
Complex source	Starring Meryl Streep, Bruce Willis, Goldie Hawn and Isabella Rossellini, the film focuses on a childish pair of rivals who drink a magic potion that promises eternal youth.
MINWIKISPLIT	(1) **The film is starring Meryl Streep.**
	(2) **The film is starring Bruce Willis.**
	(3) **The film is starring Goldie Hawn.**
	(4) **The film is starring Isabella Rossellini.**
	(5) **The film focuses on a childish pair of rivals.**
	(6) **These rivals drink a magic potion.**
	(7) **This magic potion promises eternal youth.**
Complex source	The film is a partly fictionalized presentation of the tragedy that occurred in Kasaragod District of Kerala in India, as a result of endosulfan, a pesticide used on cashew plantations owned by the government.
MINWIKISPLIT	(1) **The film is a partly fictionalized presentation of the tragedy.**
	(2) **This tragedy occurred in Kasaragod District of Kerala in India.**
	(3) **This was as a result of endosulfan.**
	(4) **Endosulfan is a pesticide.**
	(5) **This pesticide is used on cashew plantations.**
	(6) **These cashew plantations are owned by the government.**

prepositional phrases, adverbial/adjectival phrases and coordinate noun and verb phrases, participial phrases and lead noun phrases). In that way, a fine-grained output in the form of a sequence of minimal, self-contained propositions is produced. Some example instances are shown in Table 8.1.

Quality Control To ensure that the resulting dataset is of high quality, we defined a set of dependency parse and part-of-speech-based heuristics to filter out sequences that contain *grammatically incorrect* sentences, as well as sentences that *mix multiple semantic units* and, thus, are violating the specified minimality requirement. For instance, in order to verify that the simplified sentences are grammatically sound, we check whether the root node of the output sentence is a verb and whether one of its child nodes is assigned a dependency label that denotes a subject component. To test if the simplified sentences represent minimal propositions, we check whether the output does not contain a clausal modifier, such as a relative clause modifier, adverbial clause modifier or a clausal modifier of a noun. Moreover, we ensure that no conjunction is included in the simplified output sentences. In the future, we will implement some further heuristics to *avoid uniformity* in the structure of the source sentences. In that way, we aim at guaranteeing a great structural variability in the input in order to enable systems that are trained on the MINWIKISPLIT corpus to learn splitting rewrites for a wide range of linguistic constructs.

After running our sentence simplification framework DISSIM over the source sentences from the WIKISPLIT corpus and applying the set of heuristics that we defined to ensure grammaticality and minimality of the output, 203K pairs of input and corresponding output sequences were left.

Summary

9

In the prior sections, we presented a discourse-aware sentence splitting approach that creates a semantic hierarchy of syntactically simplified sentences. Our framework differs from previously proposed approaches by using a linguistically grounded transformation stage that converts complex sentences into shorter utterances with a more regular structure using clausal and phrasal disembedding mechanisms. It takes a sentence as input and performs a recursive transformation process that is based upon a small set of 35 hand-crafted grammar rules, encoding syntactic and lexical features that are derived from a sentence's phrase structure. Each rule specifies

(1) how to *split up and rephrase* the input into structurally simplified sentences (see Chapter 4), and
(2) how to *set up a contextual hierarchy* between the split components (see Section 5.1) and *identify the semantic relationship* that holds between those elements (see Section 5.2).

By disembedding clausal and phrasal constituents that contain only supplementary information, source sentences that present a complex linguistic structure are recursively split into shorter, syntactically simplified components. In that way, the input is reduced step-by-step to its key information (*"core sentence"*) and augmented with a number of associated *contextual sentences* that disclose additional information about it, resulting in a novel hierarchical representation in the form of core sentences and accompanying contexts. Moreover, we determine the rhetorical relations by which the split sentences are linked in order to preserve their semantic relationship. In that way, a semantic hierarchy of minimal propositions is established, capturing both their semantic context and relations to other units in the form of rhetorical relations.

C. Niklaus, *From Complex Sentences to a Formal Semantic Representation using Syntactic Text Simplification and Open Information Extraction*, https://doi.org/10.1007/978-3-658-38697-9_9

As a proof of concept, we developed a reference TS implementation, DISSIM, that serves as the basis on which our framework is evaluated (see Part IV).

To the best of our knowledge, this is the first time that syntactically complex sentences are *split and rephrased within the semantic context* in which they occur. In the following, we will leverage the semantic hierarchy of minimal propositions generated by our proposed TS approach to facilitate and improve the performance of Open IE applications. For this purpose, we will examine the following two assumptions:

(i) By taking advantage of the resulting fine-grained representation of complex assertions in terms of shorter sentences with a more regular structure, the complexity of downstream Open IE tasks will be reduced, thus improving their performance in terms of precision and recall.

(ii) The semantic hierarchy can be used to enrich the output of state-of-the-art Open IE systems with additional meta information, creating a novel context-preserving representation of relational tuples.

Part III
Open Information Extraction

Introduction

In the previous chapter, we presented a discourse-aware TS approach that splits and rephrases complex sentences within the semantic context in which they occur. To do so, we decompose a source sentence that presents a complex syntax into a set of self-contained propositions, with each of them presenting a minimal semantic unit. In that way, a fine-grained output with a simple and regular structure is generated. Moreover, in order to improve the interpretability of the simplified sentences, we not only split the input into isolated sentences, but also incorporate the semantic context in the form of a hierarchical structure and semantic relationships between the split propositions. In that way, complex sentences are transformed into a *semantic hierarchy of minimal propositions*, as illustrated in the upper part of Figure 10.1.

An application area that may benefit greatly from context-preserving sentence splitting as a preprocessing step is the task of Open IE, whose goal is to obtain a shallow semantic representation of large amounts of texts in the form of predicate-argument structures (see Section 2.3). While Open IE approaches have been evolving along the years to capture relations beyond simple subject-predicate-object tuples, a closer look into the extraction of the linkage between clauses and phrases within a complex sentence has been neglected. Hence, in a second step, we aim to leverage the semantic hierarchy of minimal propositions created by our discourse-aware TS approach in constructing a lightweight semantic representation of complex sentences that targets complementary linguistic phenomena that are central to the extraction of sentence-level MRs, including predicate-argument structures, clausal and phrasal linkage patterns, as well as intra-sentential rhetorical structures (see the lower part of Figure 10.1).

C. Niklaus, *From Complex Sentences to a Formal Semantic Representation using Syntactic Text Simplification and Open Information Extraction*, https://doi.org/10.1007/978-3-658-38697-9_10

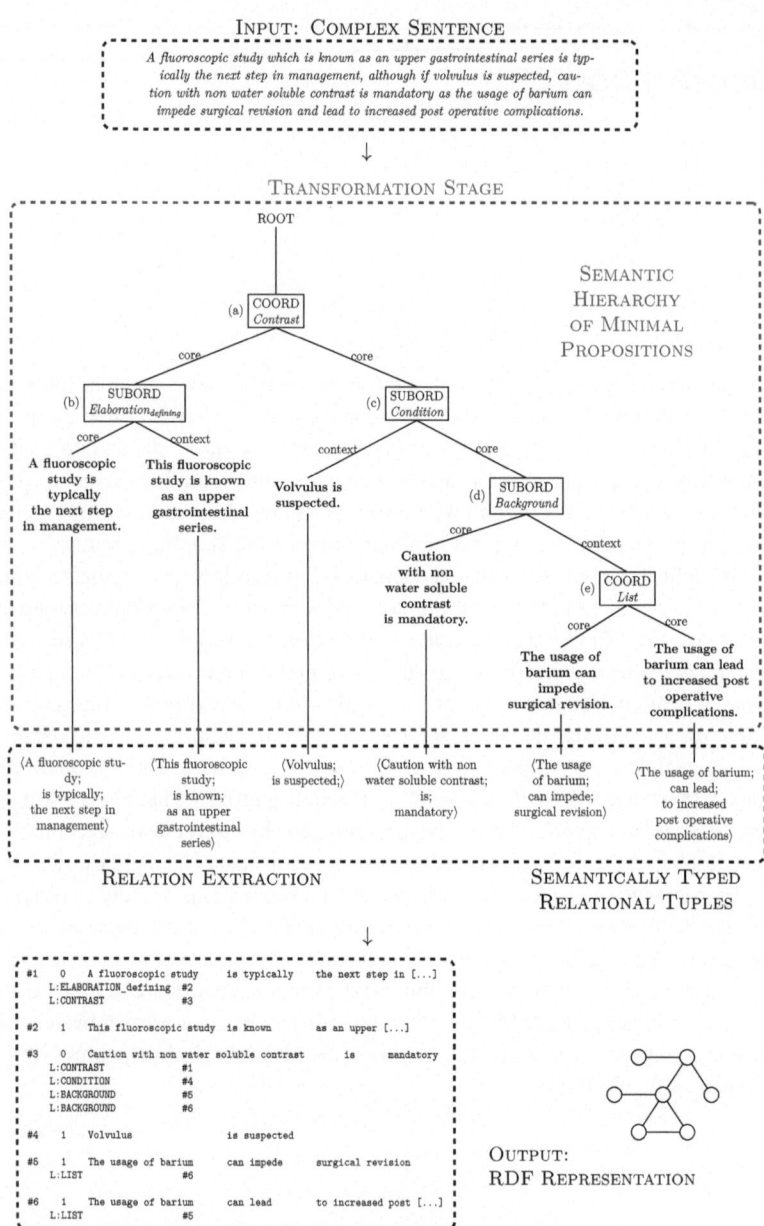

Figure 10.1 Workflow of the proposed TS—Open IE pipeline

Subtask 3: Extracting Semantically Typed Relational Tuples

11

We aim to leverage the semantic hierarchy of simplified sentences that was generated in the previous stage by our discourse-aware TS framework, using it as a pre-processing step for *extracting semantically typed relational tuples from complex input sentences.* This approach is based on the idea that the fine-grained representation of complex assertions in the form of hierarchically ordered and semantically interconnected sentences that present a simplified syntax may support the task of Open IE in two dimensions:

Coverage and Accuracy Based on the assumption that shorter sentences with a more regular structure are easier to process and analyze for downstream Open IE applications, we hypothesize that the complexity of the relation extraction task will be reduced when taking advantage of the split sentences instead of dealing directly with the raw source sentences. Hence, we expect to improve the performance of state-of-the-art Open IE systems in terms of the *precision* and *recall* of the extracted relational tuples when operating on the simplified sentences.

Context-Preserving Representation We hypothesize that the semantic hierarchy generated by our discourse-aware TS approach can be leveraged to extract relations within the *semantic context* in which they occur. By enriching the output of state-of-the-art Open IE systems with additional meta information, we intend to create a novel context-preserving representation for the task of Open IE which extends the shallow semantic representation of current approaches to produce an output that captures important contextual information. In that way, we aim to generate a representation for relational tuples that is more informative and coherent, and thus easier to interpret.

© The Author(s), under exclusive license to Springer Fachmedien Wiesbaden GmbH, 171
part of Springer Nature 2022
C. Niklaus, *From Complex Sentences to a Formal Semantic Representation using Syntactic Text Simplification and Open Information Extraction,*
https://doi.org/10.1007/978-3-658-38697-9_11

Below, we first explain how state-of-the-art Open IE approaches can make use of our discourse-aware TS approach as a pre-processing step to leverage the syntactically simplified sentences for facilitating the extraction of relational tuples and to enrich their output with semantic information (see Section 11.1). In addition, we present an Open IE reference system, Graphene, which implements a relation extraction pattern upon the semantic hierarchy of minimal propositions (see Section 11.2).

11.1   Enriching State-of-the-Art Open Information Extraction Approaches with Semantic Information

To extract a set of semantically typed relational tuples from a complex assertion, each minimal proposition $prop \in PROP$ that was generated by our discourse-aware TS approach (represented as leaf nodes in the linked proposition tree LPT) is sent as input to the Open IE system whose output shall be enriched with semantic information. For instance, in case of our running example (see Table 1.1), the input to an Open IE system are the six split sentences from Figure 7.14 (#1.1, #1.2, #2.1, #2.2.1, #2.2.2.1., #2.2.2.2). When using the state-of-the-art framework RnnOIE, the following seven relational tuples are extracted:

#1.1   ⟨A fluoroscopic study; is; typically, the next step in management⟩
#1.2   ⟨A fluoroscopic study; is known; as an upper gastrointestinal series⟩
#2.1   ⟨Caution with non water soluble contrast; is; mandatory⟩
#2.2.1   ⟨Volvulus; is suspected; ⟩
#2.2.2.1   ⟨The usage of barium; can impede; surgical revision⟩
#2.2.2.2   ⟨The usage of barium; can lead; to increased post operative complications⟩
#2.2.2.2   ⟨The usage of barium; to increased; post operative complications⟩

After identifying the predicate-argument structures of the input sentences, the extracted relational tuples need to be mapped to their semantic context. It is available from the linked proposition tree LPT in the form of semantic relationships $rel \in REL$ between the split propositions $prop \in PROP$ and their hierarchical order $c \in CL$. As Open IE systems often extract more than one relational tuple at different levels of granularity from a given simplified sentence (e.g., #2.2.2.2), we need to determine which of them represents the main statement of the underlying minimal proposition. This decision is based on the following heuristic (Cetto, 2017): an extraction is assumed to embody the main message of a sentence if

(1) the head verb of the input sentence is contained in the relational phrase *rel_phrase* of the extracted tuple (e.g., ⟨*Volvulus*; *is suspected*; ⟩); or

(2) the head verb of the input sentence equals the object argument arg_2 of the extracted tuple (e.g., ⟨*Volvulus*; *is suspected*⟩).

For example, in the proposition *"Volvulus is suspected."*, the term *"suspected"* represents the head verb. With condition (1), we are able to match tuples such as ⟨*Volvulus*; *is suspected*; ⟩, where the head verb can be found in the predicate position. Condition (2), in contrast, covers tuples that contain the head verb in object position, which is often the case for Open IE approaches that extract verbal auxiliaries (here: *"is"*) as the relational phrase and the main verb as the corresponding object argument, e.g. ⟨*Volvulus*; *is suspected*⟩.

Any extraction that matches one of the above-mentioned criteria will be denoted as a head tuple *htuple(s)* of the corresponding simplified sentence *s*. The head verbs of the minimal propositions from our running example are underlined in the extracted tuples below. By applying the heuristic described above, we determine whether or not they represent the head tuple of one of the simplified sentences. If so, we map the extraction to their corresponding minimal proposition $prop \in PROP$ from the linked proposition tree LPT.

- ⟨*A fluoroscopic study*; *is*; *typically, the next step in management*⟩
  → $htuple(\#1.1)$
- ⟨*A fluoroscopic study*; *is known*; *as an upper gastrointestinal series*⟩
  → $htuple(\#1.2)$
- ⟨*Caution with non water soluble contrast*; *is*; mandatory⟩
  → $htuple(\#2.1)$
- ⟨*Volvulus; is suspected*; ⟩
  → $htriplet(\#2.2.1)$
- ⟨*The usage of barium; can impede*; *surgical revision*⟩
  → $htuple(\#2.2.2.1)$
- ⟨*The usage of barium; can lead*; *to increased post operative complications*⟩
  → $htuple(\#2.2.2.2)$
- ⟨*The usage of barium; to increased*; *post operative complications*⟩
  →   ∅

Once the extractions are projected to the split sentences $prop \in PROP$ from the linked proposition tree LPT, the contextual hierarchy $c \in CL$ and the semantic relationship $rel \in REL$ that holds between two split sentences can be directly transferred to the extracted relational tuples, thus embedding them within the semantic context derived from the source sentence. Extractions that do not represent a head tuple, as is the case for the second extraction from sentence #2.2.2.2 in our example, are discarded.

In that way, any Open IE system can make use of our discourse-aware TS framework to enrich its extractions with semantic information. The semantically typed output produced by the state-of-the-art Open IE approach RnnOIE when applying our framework as a pre-processing step is shown in Figure 11.1 in the form of a lightweight semantic representation. For details on the output format, see Chapter 12.

To allow for a better comparison, we added the set of relational tuples that are returned by RnnOIE when directly operating on the raw source sentence (see Figure 11.2). As can be seen, in this case, its output represents only a loose arrangement of relational tuples that are hard to interpret as they ignore the context under which

```
RnnOIE (using discourse-aware TS framework):
(1) #1    0    A fluoroscopic study; is; typically, the next step in
management
(1a)            L:ELABORATION   #2
(1b)            L:CONTRAST      #3

(2) #2    1    A fluoroscopic study; is known; as an upper gastrointestinal
  series

(3) #3    0    Caution with non water soluble; is; mandatory
(3a)            L:CONTRAST      #1
(3b)            L:CONDITION     #6
(3c)            L:BACKGROUND    #4
(3d)            L:BACKGROUND    #5

(4) #4    1    The usage of barium; can impede; surgical revision
(4a)            L:LIST          #5

(5) #5    1    The usage of barium; can lead; to increased post operative
complications
(5a)            L:LIST          #4

(6) #6    1    Volvulus; is suspected;
```

Figure 11.1 Semantically enriched output of RnnOIE when using our discourse-aware TS approach as a pre-processing step. The extracted relational tuples are augmented with contextual information in terms of intra-sentential rhetorical structures and hierarchical relationships, thus preserving the coherence structure of the complex source sentence

```
RnnOIE (stand-alone):
(1) (A fluoroscopic study; known; as an upper gastrointestinal series)
(2) (caution with non water soluble contrast; is; mandatory as the usage
of barium)
(3) (as the usage; of barium can impede; surgical revision and lead)
(4) ( ; to increased; post operative complications)
```

Figure 11.2 Relational tuples extracted by RnnOIE *without* using our discourse-aware TS approach as a pre-processing step. The output represents a loose arrangement of relational tuples that are hard to interpret as they ignore the semantic context under which an extraction is complete and correct

an extraction is complete and correct, and thus lacks the expressiveness needed for a proper interpretation of the extracted predicate-argument structures. For instance, for a correct understanding of our running example, it is important to distinguish between information asserted in the sentence and information that is only conditionally true, since the information contained in the statement *"caution with non water soluble contrast is mandatory"* depends on the pre-posed if-clause. As the output depicted in Figure 11.2 shows, RnnOIE is not able to capture this relationship, though. This lack of expressiveness impedes the interpretation of the extracted tuples and is vulnerable to incorrect reasoning in downstream Open IE applications. However, with the help of the semantic hierarchy generated by our discourse-aware TS approach, its output can easily be enriched with contextual information that allows to capture the semantic relationship between the set of extracted relations and, hence, preserve their interpretability in downstream tasks, as illustrated in Figure 11.1.

Moreover, the output from Figure 11.1 demonstrates that our proposed TS approach supports the Open IE system under consideration on a second dimension: using it as a pre-processing step to split the input into a sequence of syntactically simplified sentences improves the precision and recall of the extracted relational tuples. For instance, the arguments in the object slots of extraction (2) and (3) in Figure 11.2 are inaccurate, as they incorporate extra phrasal elements that should rather form the basis of a separate extraction, thus leading to a malformed relational tuple that has no meaningful interpretation. However, when operating on the decomposed sentences, RnnOIE successfully divides the information contained in tuple (2) in Figure 11.2 into two individual extractions that are properly structured ((2) and (3) in Figure 11.1); the same holds for extraction (3) in Figure 11.2. In addition, when operating on the split components, RnnOIE succeeds in extracting the tuple ⟨*Volvulus*; *is suspected*; ⟩, which it was not able to detect when running over the original complex sentence.

11.2   Reference Implementation Graphene

In addition, we developed a reference Open IE implementation, Graphene. It implements a baseline relation extractor that extracts a binary relational tuple from a given simplified sentence (Cetto, 2017). As the phrasal parse trees of the input are available from the transformation stage of the TS process, we use a syntactic pattern that operates on a sentence's phrasal parse structure to identify subject-predicate-object tuples (see Figure 11.3).

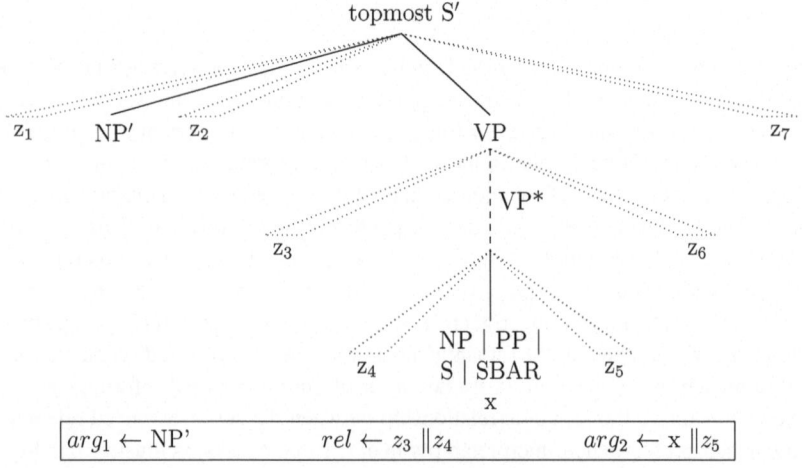

Figure 11.3 Graphene's relation extraction pattern (Cetto, 2017). "∥" denotes concatenation

For instance, when applying the extraction pattern specified in Figure 11.3 on the parse tree of the minimal proposition *"The usage of barium can impede surgical revision."* (see Figure 11.4), the following relational tuple is extracted:

⟨*The usage of barium*; *can impede*; *surgical revision*⟩

The final output of our running example, using Graphene's relation extraction implementation, is depicted in Figure 12.3.

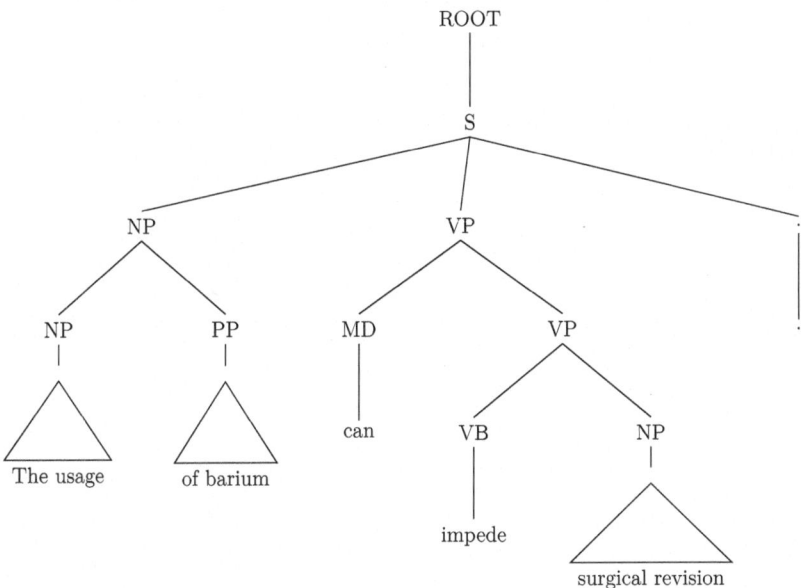

Figure 11.4 Phrasal parse tree of the minimal proposition *"The usage of barium can impede surgical revision"*

Generating a Formal Semantic Representation

12

The relational tuples that are extracted from a set of simplified sentences by our Open IE approach (see Section 11.2) are turned into a lightweight sentence-level MR in the form of semantically typed predicate-argument structures. For this purpose, we augment the shallow semantic representation of state-of-the-art Open IE systems with

(1) *rhetorical structures* that establish a semantic link between the extracted tuples, and
(2) information about the *hierarchical level* of each extraction.

12.1 Lightweight Semantic Representation for Open Information Extraction

In order to represent contextual relationships between the extracted relational tuples, the default representation of an Open IE tuple needs to be extended. Therefore, we propose a novel *lightweight semantic representation for the output of Open IE systems* that is both machine processable and human readable.

In this representation, a binary relational tuple $t \leftarrow (arg_{subj}, rel\_phrase, arg_{obj})$ is extended with (Cetto, 2017):

- a unique identifier id;
- information about the contextual hierarchy, the so-called *context layer cl*; and
- two sets of semantically classified contextual arguments C_S (*simple contextual arguments*) and C_L (*linked contextual arguments*).

© The Author(s), under exclusive license to Springer Fachmedien Wiesbaden GmbH, part of Springer Nature 2022
C. Niklaus, *From Complex Sentences to a Formal Semantic Representation using Syntactic Text Simplification and Open Information Extraction*,
https://doi.org/10.1007/978-3-658-38697-9_12

Accordingly, the tuples extracted from a complex sentence are represented as a set of (id, cl, t, C_S, C_L) tuples.

The *context layer cl* encodes the hierarchical order of core and contextual facts. Tuples with a context layer of 0 carry the core information of a sentence, whereas tuples with a context layer of $cl > 0$ provide contextual information about extractions with a context layer of $cl - 1$. Both types of contextual arguments, C_S and C_L, provide (semantically classified) contextual information about the statement expressed in t. Whereas a simple contextual argument $c_S \in C_S, c_S \leftarrow (s, rel)$ contains a textual span s that is classified by the semantic relation rel, a linked contextual argument $c_L \in C_L, c_L \leftarrow (id(z), rel)$ refers to the content expressed in another relational tuple z.

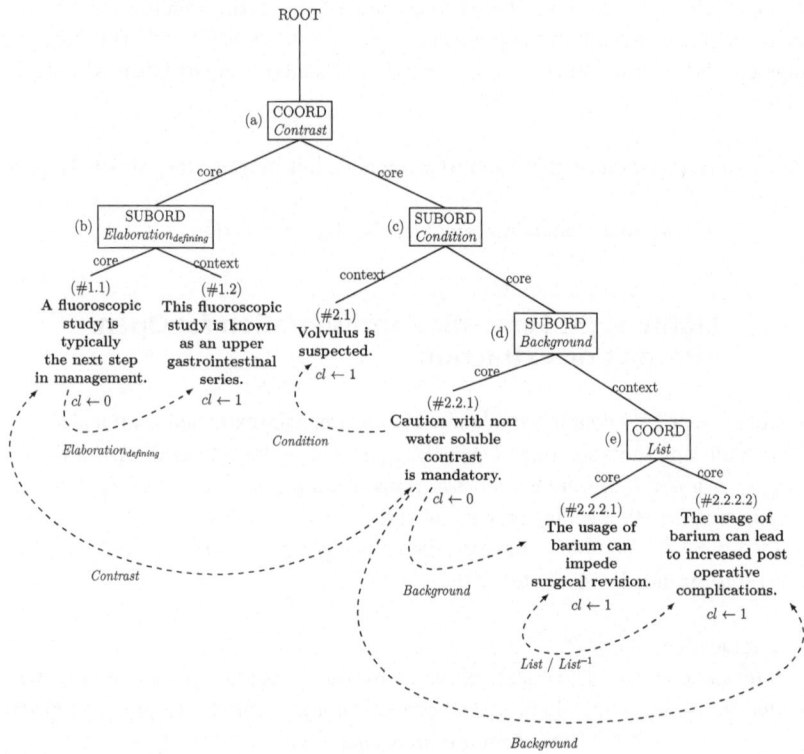

Figure 12.1 Final linked proposition tree of our example sentence

Example In the following, we will illustrate how the complex sentence from our running example from Table 1.1, whose semantic hierarchy is depicted in Figure 12.1 in the form of a linked proposition tree, is transformed into the lightweight semantic representation described above.

In the relation extraction stage, the head tuple t is extracted from each leaf node of the linked proposition tree, i.e. from each simplified sentence. In our case, this process results in the following predicate-argument tuples:

- t_1: ⟨*A fluoroscopic study; is typically; the next step in management*⟩
- t_2: ⟨*This fluoroscopic study; is known; as an upper gastrointestinal series*⟩
- t_3: ⟨*Volvulus; is suspected; *⟩
- t_4: ⟨*Caution with non water soluble contrast; is; mandatory*⟩
- t_5: ⟨*The usage of barium; can impede; surgical revision*⟩
- t_6: ⟨*The usage of barium; can lead; to increased post operative complications*⟩

Next, a tuple (id, cl, t, C_S, C_L) is created for every extraction as follows (Cetto, 2017):

The value of the context layer cl is determined by counting the number of context edges in the path from the root node to the corresponding simplified sentence in the linked proposition tree. Our approach is based on the assumption that the core information of any internal node n in the tree is represented by sentences that

(1) are leaf nodes of the subtree spanned by n, and
(2) are connected to n by a path that consists of core edges only.

Such a path is referred to as a *core chain*. The set of leaf nodes that are connected to n by a core chain are called *representatives* of n. For instance, the core information of node (a) in Figure 12.1 can be found in the representatives (#1.1) and (#2.2.1) by following the core chains that pass through the nodes (a)(b)(#1.1) and (a)(c)(d)(#2.2.1). We further claim that any relationship rel (represented by an internal node n) that holds between the child nodes of n (let a and b denote these child nodes in the binary setting) also holds pairwisely between representatives of a and b. Figure 12.2 illustrates an example. As visualized by the dashed arrows, the relation rel between the nodes a and b applies to the following pairs of sentences: $\{\{1, 3\}, \{1, 4\}, \{2, 3\}, \{2, 4\}\}$. Note that these arrows define a direction for each relation rel that is established between two sentences x and y. Such a directed edge (x, y) is called a *contextual link* for the relation rel in our approach.

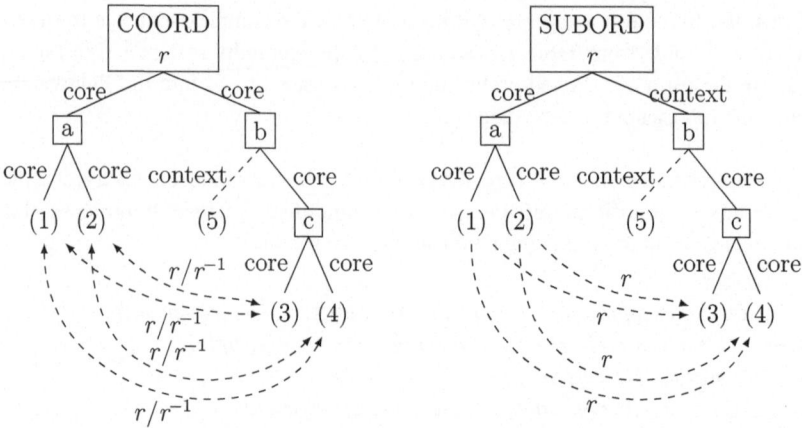

Figure 12.2 Establishing contextual links between sentences connected by coordination nodes *(left)* and subordination nodes *(right)* (Cetto, 2017)

The direction of the contextual link indicates how the relationship between the two sentences is represented: for each contextual link (x, y), we will attach y to x as a contextual argument, semantically classified by the corresponding relation rel. By default, the identifier of y will be added to the set of linked contextual arguments C_L of the sentence x.

The contextual argument assignment depends on the constituency type of the corresponding relational node, as shown in Figure 12.2. For a coordination node n, we create two contextual links for each relation rel that holds between two representatives x and y: (x, y) for the relation rel and (y, x) for the inverse relation rel^{-1} (see Table 5.1a). Consequently, core sentence representatives x and y are mutually connected to each other via linked contextual arguments classified as rel and rel^{-1}. In contrast, subordination nodes n establish a single contextual link (x, y) for each relation rel that holds between two representatives x and y, with y being inside the contextual subtree of n. As a result, a core sentence representative x is enriched with a linked contextual argument that refers to the context sentence representative y, but not vice versa.

The final output of our running example in the form of our proposed lightweight semantic representation, using Graphene's relation extraction implementation, is shown in Figure 12.3.

```
(1) #1    0    A fluoroscopic study;  is typically;    the next step in
management
(1a)                L:ELABORATION_defining #2
(1b)                L:CONTRAST              #3


(2) #2    1    This fluoroscopic study;  is known;        as an upper
gastrointestinal series


(3) #3    0    Caution with non water soluble contrast;  is;   mandatory
(3a)                L:CONTRAST             #1
(3b)                L:CONDITION            #4
(3c)                L:BACKGROUND           #5
(3d)                L:BACKGROUND           #6


(4) #4    1    Volvulus;                is suspected;


(5) #5    1    The usage of barium;   can impede;    surgical revision
(5a)                L:LIST                 #6


(6) #6    1    The usage of barium;   can lead;      to increased post
operative complications
(6a)                L:LIST                 #5
```

Figure 12.3 Human readable representation (RDF-NL) of our running example (when using Grapene's relation extraction implementation)

12.2   Human Readable Representation

To facilitate a manual analysis of the extracted relational tuples, a human readable format, called RDF-NL, is generated by our framework (see Figure 12.3 and Figure 12.4) (Cetto, 2017). In this format, relational tuples are represented by tab-separated strings, specifying the identifier id, context layer cl and the core extraction (e.g., (1) and (2) in Figure 12.3) that is represented by a binary relational tuple $t \leftarrow (arg_{subj}, rel\_phrase, arg_{obj})$:

- subject argument arg_{subj},
- relation name rel, and
- object argument arg_{obj}.

Contextual arguments (C_S and C_L) are indicated by an extra indentation level to their parent tuples (e.g., (1a) and (1b) in Figure 12.3). Their representation consists of a type string and a tab-separated content. The type string encodes both the context type (S for a simple contextual argument $c_S \in C_S$ and L for a linked contextual argument $c_L \in C_L$) and the classified semantic relation (e.g., *Cause*, *Purpose*). The content of a simple contextual argument is a single phrase (e.g., (4a) in Figure 12.4), whereas the content of a linked contextual argument is the identifier of the respective target extraction, which again represents an entire relational tuple (e.g., (4b) in Figure 12.4).

```
(1) #1   0   The house;      was once;    part of a plantation
(1a)             L:LIST                       #2


(2) #2   0   It;             was;         the home of Josiah Henson
(2a)             L:ELABORATION_non_defining #3
(2b)             L:LIST                       #1


(3) #3   1   Josiah Henson; was;          a slave
(3a)             L:ELABORATION_defining      #4
(3b)             L:ELABORATION_defining      #5


(4) #4   2   A slave;        escaped;     to Canada
(4a)             S:TEMPORAL                   in 1830
(4b)             L:LIST                       #5


(5) #5   2   A slave;        wrote;       the story of his life
(5a)             L:LIST                       #4
```

Figure 12.4 Human readable representation (RDF-NL) of the sentence *"The house was once part of a plantation and it was the home of Josiah Henson, a slave who escaped to Canada in 1830 and wrote the story of his life."* (when using Grapene's relation extraction implementation)

12.3   Machine Readable Representation

In addition, our framework can materialize its extractions into a graph serialized under the N-Triples[1] specification of the RDF standard (Cetto, 2017), facilitating machine processing of the extracted relations in downstream applications.

The machine readable output of our running example is shown in Figure 12.5. In this format, each sentence is modeled as a blank node comprising the original sentence (1) and a list of contained extractions ((2) to (13)). Each relational tuple is represented as a blank node of RDF text resources that represent the subject argument arg_{subj} (6), the relation name rel (7) and the object argument arg_{obj} (8) of the relational tuple $t \leftarrow (arg_{subj}, rel\_phrase, arg_{obj})$. Information about the context layer cl (5) is stored as well. Contextual information is attached by RDF predicates that encode both the context type and the classified semantic relation, analogous to the human readable format. These RDF predicates link to either text resources in the case of a simple contextual argument[2] or to other extractions (9, 11) in the case of a linked contextual argument. Finally, all RDF text resources are mapped to their unescaped literal values (12).

---

[1] https://www.w3.org/TR/n-triples

[2] An example of a text resource denoting a simple contextual argument can be found in (8) in Figure 12.6.

A fluoroscopic study which is known as an upper gastrointestinal series is
typically the next step in management, although if volvulus is suspected,
caution with non water soluble contrast is mandatory as the usage of
barium can impede surgical revision and lead to increased post operative
complications.

```
(1)     _:fa89da284aee42469dafe015bb2b79e1
<http://lambda3.org/graphene/sentence#original-text>
"A fluoroscopic study which is known as an upper gastrointestinal series
is typically the next step in management , although if volvulus is
suspected , caution with non water soluble contrast is mandatory as the
usage of barium can impede surgical revision and lead to increased post
operative complications
."^^<http://www.w3.org/2001/XMLSchema#string> .
(2)     _:fa89da284aee42469dafe015bb2b79e1
        <http://lambda3.org/graphene/sentence#has-extraction>
(3)     _:7a43269ddb2a4489a76256278c2534df .
(4)     _:7a43269ddb2a4489a76256278c2534df
        <http://lambda3.org/graphene/extraction#extraction-type>
        "VERB_BASED"^^<http://www.w3.org/2001/XMLSchema#string> .
(5)     _:7a43269ddb2a4489a76256278c2534df
        <http://lambda3.org/graphene/extraction#context-layer>
        "0"^^<http://www.w3.org/2001/XMLSchema#integer> .
(6)     _:7a43269ddb2a4489a76256278c2534df
        <http://lambda3.org/graphene/extraction#subject>
        <http://lambda3.org/graphene/text#A+fluoroscopic+study> .
(7)     _:7a43269ddb2a4489a76256278c2534df
        <http://lambda3.org/graphene/extraction#predicate>
        <http://lambda3.org/graphene/text#is+typically> .
(8)     _:7a43269ddb2a4489a76256278c2534df
        <http://lambda3.org/graphene/extraction#object>
        <http://lambda3.org/graphene/text#the+next+step+in+management> .
(9)     _:7a43269ddb2a4489a76256278c2534df
        <http://lambda3.org/graphene/extraction#L-IDENTIFYING_DEFINITION>
(10)    _:27c44ff725af4dcd81b78d7026a464e9 .
(11)    _:7a43269ddb2a4489a76256278c2534df
        <http://lambda3.org/graphene/extraction#L-CONTRAST>
(12)    _:cb9dd041b06e4001b6a2b3947cdeb688 .
        <http://lambda3.org/graphene/text#A+fluoroscopic+study>
        <http://www.w3.org/1999/02/22-rdf-syntax-ns#value> "A fluoroscopic
        study"^^<http://www.w3.org/2001/XMLSchema#string> .
        <http://lambda3.org/graphene/text#is+typically>
        <http://www.w3.org/1999/02/22-rdf-syntax-ns#value> "is
        typically"^^<http://www.w3.org/2001/XMLSchema#string> .
        <http://lambda3.org/graphene/text#the+next+step+in+management>
        <http://www.w3.org/1999/02/22-rdf-syntax-ns#value> "the next step in
        management"^^<http://www.w3.org/2001/XMLSchema#string> .
(13)    _:fa89da284aee42469dafe015bb2b79e1
        <http://lambda3.org/graphene/sentence#has-extraction>
...
```

Figure 12.5 Machine readable representation (RDF-Triples) of our running example (when
using Grapene's relation extraction implementation)

The house was once part of a plantation and it was the home of Josiah Henson,
a slave who escaped to Canada in 1830 and wrote the story of his life.

(1)   _:206b3c28cdad40df8b995b05bad3d8ac
        <http://lambda3.org/graphene/sentence#original-text>
        "The house was once part of a plantation and it was the
        home of Josiah Henson , a slave who escaped to Canada
        in 1830 and wrote the story of his life
        ."^^<http://www.w3.org/2001/XMLSchema#string> .
...

(2)   _:206b3c28cdad40df8b995b05bad3d8ac
        <http://lambda3.org/graphene/sentence#has-extraction>
        _:ae87562d67c5425ba321927796091c74 .

(3)   _:ae87562d67c5425ba321927796091c74
        <http://lambda3.org/graphene/extraction#extraction-type>
        "VERB_BASED"^^<http://www.w3.org/2001/XMLSchema#string> .

(4)   _:ae87562d67c5425ba321927796091c74
        <http://lambda3.org/graphene/extraction#context-layer>
        "2"^^<http://www.w3.org/2001/XMLSchema#integer> .

(5)   _:ae87562d67c5425ba321927796091c74
        <http://lambda3.org/graphene/extraction#subject>
        <http://lambda3.org/graphene/text#A+slave> .

(6)   _:ae87562d67c5425ba321927796091c74
        <http://lambda3.org/graphene/extraction#predicate>
        <http://lambda3.org/graphene/text#escaped> .

(7)   _:ae87562d67c5425ba321927796091c74
        <http://lambda3.org/graphene/extraction#object>
        <http://lambda3.org/graphene/text#to+Canada> .

(8)   _:ae87562d67c5425ba321927796091c74
        <http://lambda3.org/graphene/extraction#S-TEMPORAL>
        <http://lambda3.org/graphene/text#in+1830+.> .

(9)   _:ae87562d67c5425ba321927796091c74
        <http://lambda3.org/graphene/extraction#L-LIST>
        _:ee5ea62c09664638801cf665aefdfeb0 .

(10)   <http://lambda3.org/graphene/text#A+slave>
        <http://www.w3.org/1999/02/22-rdf-syntax-ns#value> "A
        slave"^^<http://www.w3.org/2001/XMLSchema#string> .

(11)   <http://lambda3.org/graphene/text#escaped>
        <http://www.w3.org/1999/02/22-rdf-syntax-ns#value>
        "escaped"^^<http://www.w3.org/2001/XMLSchema#string> .

(12)   <http://lambda3.org/graphene/text#to+Canada>
        <http://www.w3.org/1999/02/22-rdf-syntax-ns#value> "to
        Canada"^^<http://www.w3.org/2001/XMLSchema#string> .

(13)   <http://lambda3.org/graphene/text#in+1830+.>
        <http://www.w3.org/1999/02/22-rdf-syntax-ns#value> "in 1830
        ."^^<http://www.w3.org/2001/XMLSchema#string> .
...

Figure 12.6 Machine readable representation (RDF-Triples) of the example sentence from
Figure 12.4 (when using Grapene's relation extraction implementation)

Summary

13

Previous work in the area of Open IE has mainly focused on the extraction of isolated relational tuples. Ignoring the cohesive nature of texts where important contextual information is spread across clauses or sentences, state-of-the-art Open IE approaches are prone to generating a loose arrangement of tuples that lack the expressiveness needed to infer the true meaning of complex assertions.

To overcome this limitation, we presented a method that allows existing Open IE systems to enrich their output with semantic information. By leveraging the semantic hierarchy of minimal propositions generated by our discourse-aware TS approach, they are able to extract semantically typed relational tuples from complex source sentences. In that way, the semantic context of the input sentences is preserved, resulting in more informative and coherent predicate-argument structures which are easier to interpret.

Based on this novel type of output in the form of semantically typed relational tuples, we proposed a lightweight semantic representation for Open IE. It extends the shallow semantic representation of state-of-the-art approaches in the form of predicate-argument structures by capturing intra-sentential rhetorical structures and hierarchical relationships between the relational tuples. In that way, the semantic context of the extracted tuples is preserved, allowing for a proper interpretation of the output.

The fine-grained representation of sentences in the form of minimal propositions that present a simplified syntax may support the task of Open IE on a second dimension. Since the simplified sentences present a more regular structure, they are easier to process and analyze. Therefore, we expect to improve the coverage and accuracy of the tuples extracted by state-of-the-art Open IE approaches when taking advantage of the split sentences instead of dealing with the raw source sentences. This hypothesis will be examined in detail in Section 15.3.2.1.

© The Author(s), under exclusive license to Springer Fachmedien Wiesbaden GmbH, part of Springer Nature 2022
C. Niklaus, *From Complex Sentences to a Formal Semantic Representation using Syntactic Text Simplification and Open Information Extraction*,
https://doi.org/10.1007/978-3-658-38697-9_13

189

Besides of supporting existing Open IE approaches as a pre-processing step, we also developed a reference Open IE implementation, Graphene. It implements a baseline relation extractor on top of the hierarchically ordered and semantically interconnected sentences that extracts a binary relational tuple from a given simplified sentence.

Part IV
Evaluation

Experimental Setup

14

In the following, we present the experimental setup for evaluating the performance of our proposed discourse-aware TS approach with regard to the two subtasks of

(1) splitting and rephrasing syntactically complex input sentences into a set of *minimal propositions* (**Hypotheses 1.1, 1.2 and 1.3**), and
(2) setting up a *semantic hierarchy* between the split components, based on the tasks of constituency type classification and rhetorical relation identification (**Hypotheses 1.4 and 1.5**).

Moreover, we describe the setting for assessing the use of discourse-aware sentence splitting as a support for the task of Open IE (**Hypotheses 2.1 and 2.2**).

14.1 Subtask 1: Splitting into Minimal Propositions

In a first step, we focus on evaluating the first subtask, whose goal is to split sentences that present a complex syntax into minimal, self-contained sentences with a simplified structure.

Supplementary Information The online version contains supplementary material available at https://doi.org/10.1007/978-3-658-38697-9_14.

C. Niklaus, *From Complex Sentences to a Formal Semantic Representation using Syntactic Text Simplification and Open Information Extraction*,
https://doi.org/10.1007/978-3-658-38697-9_14

14.1.1 Datasets

To evaluate the performance of our discourse-aware TS framework with regard to the sentence splitting subtask, we conducted experiments on three commonly applied TS corpora from two different domains. The first dataset we used was *WikiLarge*. Its test set consists of 359 sentences from the PWKP corpus, which is made up of aligned EW sentences and SEW counterparts. In addition, we assessed the performance of our simplification system on the 5,000 test sentences from the *WikiSplit* benchmark, which was mined from Wikipedia edit histories. Moreover, to demonstrate domain independence, we compared the output generated by our TS approach with that of the various baseline systems on the *Newsela* corpus, whose test set is composed of 1,077 sentences from newswire articles.

14.1.2 Baselines

We compared our reference implementation DISSIM against several state-of-the-art TS baseline systems that have a strong focus on syntactic transformations through explicitly modeling splitting operations. For WikiLarge, these include

  (i) the recent semantic-based DSS model that decomposes a sentence into its main semantic constituents using a semantic parser;
 (ii) SENTS, which is an extension of DSS that runs the split sentences through the NTS system proposed in Nisioi et al. (2017);
(iii) HYBRID, a further TS approach that takes semantically-shared elements as the basis for splitting and rephrasing a sentence;
 (iv) the hybrid TS system YATS; and
  (v) RegenT, another hybrid state-of-the-art simplification approach for combined lexical and syntactic simplification.

In addition, we report evaluation scores for both the complex input sentences, which allows for a better judgment of system conservatism, and the corresponding simple reference sentences.

With respect to the Newsela dataset, we considered the same baseline systems, with the exceptions of DSS and SENTS, whose outputs were not available. Finally, regarding the WikiSplit corpus, we restricted the comparison of our TS approach to the best-performing system in Botha et al. (2018), Copy512, which is a Seq2Seq neural model augmented with a copy mechanism and trained over the WikiSplit dataset.

14.1.3 Automatic Metrics

Firstly, we assessed the systems' performance on the sentence splitting subtask in a comparative analysis, using a set of automatic evaluation metrics.

14.1.3.1 Descriptive Statistics
The automatic metrics that were calculated in the evaluation procedure comprise a number of descriptive statistics for better understanding the generated output, including

   (i)   the average sentence length of the simplified sentences in terms of the average number of tokens per output sentence (#T/S);
  (ii)   the average number of simplified output sentences per complex input (#S/C);
 (iii)   the percentage of sentences that are copied from the source without performing any simplification operation (%SAME), serving as an indicator for system conservatism, i.e. the tendency of a system to retain the input sentence rather than transforming it; and
  (iv)   the averaged word-based Levenshtein distance from the input (LD_{SC}), which provides further evidence for a system's conservatism.

14.1.3.2 Translation Assessment
In accordance with prior work on TS, we report average BLEU (Papineni et al., 2002) and SARI (Xu et al., 2016) scores for the rephrasings of each system. BLEU (**Bi**Lingual **E**valuation **U**nderstudy) measures the n-gram precision of a system's output against one or more references. This metric was designed for the evaluation of bilingual translation tasks, but it is also widely used for assessing the quality of monolingual translations, such as TS. BLEU ignores recall, for which it compensates by using a brevity penalty. Accordingly, it favors an output that is not too short and contains only n-grams that occur in at least one of the references.

While BLEU is based solely on the similarity of the output with the references, SARI (**S**ystem output **A**gainst **R**eferences and **I**nput sentence) considers both the input and the references in its calculation, taking into account both precision and recall. It compares the n-grams of a system's output with those of the input and the references, rewarding outputs that modify the input in ways that are expressed by the references, i.e. by *adding* n-grams that occur in any of the references but not in the input, by *keeping* n-grams that appear both in the input and the references, and by not over-*deleting* n-grams. However, a major disadvantage of SARI is the limited number of simplification transformations taken into account. Focusing on determining the quality of the words that are added, deleted and kept, it is well suited to the evaluation of lexical and simple syntactic paraphrasing operations, such as deletion and insertion, rather than more complex transformations like sentence splitting.[1]

14.1.3.3 Syntactic Complexity

We computed the SAMSA (Simplification Automatic evaluation Measure through Semantic Annotation) and $SAMSA_{abl}$ scores of each system. They are the first metrics that explicitly target syntactic aspects of TS (Sulem et al., 2018b). Unlike BLEU and SARI, these measures do not require simplification references.

The SAMSA metric is based on the idea that an optimal split of the input is one where each predicate-argument structure is assigned its own sentence in the simplified output and measures to what extent this assertion holds for the input-output pair under consideration. Accordingly, the SAMSA score is maximized when each split sentence represents exactly one semantic unit from the input. $SAMSA_{abl}$, in contrast, does not penalize cases where the number of sentences in the simplified output is lower than the number of events contained in the input, indicating separate semantic units that should be split into individual target sentences for obtaining minimal propositions.

14.1.3.4 Rule Application Statistics

To get an idea about how often each of the 35 linguistically principled transformation patterns is fired by our reference TS implementation DISSIM during the simplification of the sentences from the WikiLarge, Newsela and WikiSplit corpus, we examined the frequency distribution of rule triggerings for each syntactic construct that is addressed by our TS approach (see Table 4.2). The results provide insights into the relative relevance of the specified grammar rules, as well as their coverage with respect to syntactic phenomena involved in complex sentence constructions.

14.1.4 Manual Analyses

Even though costly to produce, human evaluation is still the preferred method for assessing the quality of the generated simplified output in TS tasks (Alva-Manchego et al., 2020). Therefore, besides comparing the output of our proposed TS approach with the above-mentioned baseline systems by means of automatic measures, providing fast and cheap results, we resort to a human evaluation for a more detailed analysis of the simplified output.

---

[1] For a detailed analysis, see Section 21 in the online supplemental material.

14.1.4.1 Human Judgments

The most widely recognized evaluation methodology to determine the quality of a simplification draws on human judges to rate pairs of complex source and simplified output sentences according to the three criteria of grammaticality, meaning preservation and simplicity, using a Likert scale that ranges from 1 to 5. Hence, to assess how good the simplification output of our proposed TS approach is compared to the baseline systems, human evaluation was carried out on a subset of 50 randomly sampled sentences per corpus by two in-house non-native, but fluent English speakers who rated each input-output pair for the different systems according to three parameters: grammaticality, meaning preservation and structural simplicity.

The detailed guidelines presented to the annotators are given in Table 14.1. Regarding the grammaticality and meaning preservation dimensions, we adopted the guidelines from Štajner and Glavaš (2017), with some minor deviations to better reflect our goal of simplifying the *structure* of the input sentences, while *maintaining their full informational content*. Besides, since the focus of our work is on structural rather than lexical simplification, we followed the approach taken in Sulem et al. (2018c) in terms of the third parameter, simplicity, and neglected the lexical simplicity of the output sentences. Instead, we restricted our analysis to the syntactic complexity of the resulting sentences, which was measured on a scale that ranges from -2 to 2 in accordance with Nisioi et al. (2017).

14.1.4.2 Qualitative Analysis of the Transformation Patterns

In order to get further insights into the quality of our implemented simplification patterns, we performed an extensive qualitative analysis of the 35 hand-crafted transformation rules, including a manual analysis of the simplification patterns that we defined, as well as a detailed error analysis. For this purpose, we compiled a dataset consisting of 100 Wikipedia sentences per syntactic phenomenon tackled by our TS approach (see Table 4.2).[2] In the construction of this corpus we ensured that the collected sentences exhibit a great syntactic variability to allow for a reliable predication about the coverage and accuracy of the specified transformation rules.

---

[2] available under https://github.com/Lambda-3/DiscourseSimplification/blob/master/ supplemental_material/dataset_pattern_analysis.zip

Table 14.1 Annotation guidelines for the human evaluation task

1. **Grammaticality (G): Is the output fluent and grammatical?**

   5 The output is meaningful and there are no grammatical mistakes.

   4 One or two minor errors, but the meaning can be easily understood.

   3 Several errors, but it is still possible to understand the meaning.

   2 It is hard to understand the meaning due to many grammatical errors.

   1 The output is so ungrammatical that it is impossible to infer any meaning.

2. **Meaning Preservation (M): Does the output preserve the meaning of the input?**

   5 The simplification has the same meaning as the original sentence. No piece of information from the input is lost.

   4 1. The core meaning is the same as in the original sentence but with some subtle differences.

       2. If there are several output sentences, one is missing some important piece of information from the original sentence but the others fully preserve the original meaning.

   3 1. The simplified sentence(s) contain(s) a part of the relevant information from the original sentence, but another important part of relevant information is missing (there is no spurious information).

       2. Incorrect pronoun/attachment resolution.

   2 The simplification has the opposite or very different meaning compared to the original sentence.

   1 Simplified sentence(s) have no coherent meaning (i.e., it is not possible to compare it/them to the original sentence).

3. **Structural Simplicity (S): Is the output simpler than the input,** *ignoring the complexity of the words?*

   2 The simplified output is *much simpler* than the original sentence.

   1 The simplified output is *somewhat simpler* than the original sentence.

   0 The simplified output is *equally difficult* as the original sentence.

   -1 The simplified output is *somewhat more difficult* than the original sentence.

   -2 The simplified output is *much more difficult* than the original sentence.

14.2   Subtask 2: Establishing a Semantic Hierarchy

In a second step, we concentrate on assessing the quality of the semantic hierarchy that is established between the split sentences by our discourse-aware TS approach. For this purpose, we conduct both an automatic and a human evaluation to examine whether

(1) the contextual hierarchy constituted between two decomposed spans is correct, i.e. accurately distinguishing *core* sentences that contain the key information of the input from *contextual* sentences that disclose less relevant, supplementary information; and
(2) the *semantic relationship* that holds between a pair of split sentences is correctly identified and classified, thus preserving the coherence structure of complex sentences in the simplified output.

14.2.1   Automatic Metrics

We automatically evaluate the constituency type classification and rhetorical relation identification steps by mapping the simplified sentences that were generated in the sentence splitting subtask to the EDUs of the RST-DT corpus,[3] a collection of 385 Wall Street Journal articles from the Penn Treebank annotated with rhetorical relations based on the RST framework.

In RST, a text's discourse structure is represented as a tree where

1. the leaves correspond to *EDUs*;
2. the internal nodes of the tree correspond to *contiguous text spans*;
3. each node is characterized by its *nuclearity*; and
4. each node is also characterized by a *rhetorical relation* between two or more non-overlapping adjacent text spans.

On average, each document contained in the RST-DT consists of 458 words and 57 EDUs. In total, the corpus comprises 21,789 discourse units, averaging to 8.1 words each.[4]

For matching simplified sentences generated by our discourse-aware TS approach to the annotations of the RST-DT corpus, we compare each split sentence to all the

---

[3] https://catalog.ldc.upenn.edu/LDC2002T07
[4] https://www.isi.edu/~marcu/discourse/Corpora.html

EDUs of the corresponding input sentence. For each pair, we search for the longest contiguous matching subsequence. Next, based on the size of the matched sequences, a similarity score between the two input strings is calculated.[5] Each pair whose similarity score surpasses an empirically determined threshold of 0.65 is considered a match.

14.2.1.1 Constituency Type Classification

To determine whether the hierarchical relationship that was assigned by our TS framework between a pair of simplified sentences is correct, we check if the hierarchy of its contextual layers corresponds to the nuclearity of the aligned text fragments of the RST-DT. For this purpose, we make use of the nuclearity status encoded in the annotations of this dataset. In addition, we compare the performance of our TS approach with that of a set of widely used sentence-level discourse parsers on this task.

14.2.1.2 Rhetorical Relation Identification

To assess the performance of the rhetorical relation identification step, we first determine the distribution of the relation types allocated by our discourse-aware TS approach when operating on the 7,284 input sentences of the RST-DT and compare it to the distribution of the manually annotated rhetorical relations of this corpus. In a second step, we examine for each matching sentence pair whether the rhetorical relation assigned by our TS framework equates the relation that connects the corresponding EDUs in the RST-DT dataset. For this purpose, we apply the more coarse-grained classification scheme from Taboada and Das (2013), who group the full set of 78 rhetorical relations that are used in the RST-DT corpus into 19 classes of relations that share rhetorical meaning.[6] Finally, we analyze the performance of our reference TS implementation DISSIM on the relation labeling task in comparison to a number of discourse parser baselines.

---

[5] In our implementation, we use the "`SequenceMatcher`" class from the python module "`difflib`".

[6] The 19 classes of rhetorical relations can be found in Table 15.17. Most importantly, Taboada and Das (2013) map the *List* and *Disjunction* relationships to a common *Joint* class and integrate the *Result* relation into the semantic category of *Cause* relationships. Moreover, both *Temporal-After* and *Temporal-Before* become part of the *Temporal* class of rhetorical relations.

14.2.2 Manual Analysis

To get a deeper insight into the accuracy of the semantic hierarchy established between the split components, the automatic evaluation described above is complemented by a manual analysis, in which three human judges independently of each other assessed each decomposed sentence according to the following four criteria:

Limitation to core information *Is the simplified output limited to core information of the input sentence, i.e. does it not contain any background information?* *(yes—no—malformed)* One of the main goals of our discourse-aware TS approach is to separate key information from less relevant secondary material expressed in the input. For this purpose, we distinguish core propositions from contextual sentences, which are indicated by a context layer of level one and above ($cl > 0$) in our reference implementation DISSIM. Contrary to core sentences ($cl = 0$), decomposed spans that are labelled as contextual sentences are not assumed to provide a central piece of information from the source. Therefore, contextual sentences are not considered with regard to this category. All other output sentences are scored, receiving a positive mark ("yes") if they are restricted to core information from the source sentence, otherwise they get a negative tag ("no"). To filter out malformed sentences, we further added a third scoring option ("malformed") that allows for marking up inaccurate sentences, i.e. sentences that do not have a meaningful interpretation, that lack critical information or are not asserted by the source.[7]

Soundness of the contextual proposition *Does the simplified sentence express a meaningful context fact?* *(yes—no)* Analogous to the core proposition criterion evaluated above, we focus upon the contextual sentences ($cl > 0$) in the following. Contextual propositions are supposed to further specify the information provided in their respective parent sentence, conveying some additional background data. This question aims to judge the soundness of each contextual proposition, flagging it with a positive mark if the simplified sentence is informative, has a meaningful interpretation and is asserted by the input. In case of a malformed output, it receives a "no" tag.[8]

---

[7] For the sake of simplicity, we do not judge the output for minimality here.

[8] In a pre-study with one annotator we further examined whether the content of a simplified sentence that is flagged as a contextual proposition by our TS approach is indeed limited to background information. This analysis revealed that only 0.5% of them were misclassified, i.e. they convey some piece of key information from the input, while 90.9% were correctly classified as context sentences. Instead, we figured out that malformed output is a bigger issue (9.6%). Therefore, we decided to not further investigate the question of limitation to

Correctness of the context allocation *Is the contextual sentence assigned to the parent sentence to which it refers? (yes—no)* To preserve the original meaning of a sentence, it is important that the contextual proposition is assigned to the correct parent sentence, i.e. the span whose content it further specifies. Otherwise, information is lost or even distorted. Hence, with this score, we estimate the correctness of this allocation by selecting either a positive ("yes") or a negative flag ("no").

Properness of the identified semantic relationship *Is the contextual sentence linked to its parent sentence via the correct semantic relation? (yes—no—unspecified)* Finally, we examine whether our proposed discourse-aware TS approach succeeds in establishing the correct semantic relationship between the context sentence under consideration and its allocated parent sentence. For this purpose, the annotators can choose between the following three alternatives: "yes", "no" and "unspecified". The third option allows to denote all the cases where our TS framework was not able to identify a rhetorical relation that holds between a given sentence pair.

The first three categories of our analysis address the correctness of the constituency type classification task (subtask 2a), while the latter targets the rhetorical relation identification step (subtask 2b). The annotation task was carried out on the simplified sentences generated by our reference TS implementation DISSIM on a random sample of 100 sentences from the OIE2016 Open IE benchmark (see Section 14.3.2). The returned output comprises 135 propositions that were flagged as core sentences and 306 propositions presenting contextual information.

14.3   Subtask 3: Extracting Relations and their Arguments

Assuming that the fine-grained representation of complex sentences in the form of hierarchically ordered and semantically interconnected sentences that present a simplified syntax supports the extraction of semantically typed relational tuples from complex input sentences, we investigate the use of our discourse-aware TS approach for downstream Open IE applications in an extrinsic evaluation. For this purpose, we integrate our TS framework as a pre-processing step into a variety of state-of-the-art Open IE systems, including REVERB, OLLIE, ClausIE, Stanford Open IE, PropS, OpenIE-4, MinIE, OpenIE-5 and RnnOIE (see Section 14.3.1). We

---

background information, and rather focus on the soundness of the contextual propositions in our human evaluation.

then determine whether these systems profit from our proposed discourse-aware TS approach along two different dimensions:

(1) In a first step, we measure if their performance is improved in terms of *recall* and *precision* of the extracted relational tuples (see Section 14.3.3.2).

(2) In a second step, we examine whether the semantic hierarchy generated by our approach is beneficial in *extending the shallow semantic representation of state-of-the-art Open IE approaches with additional meta information* to produce an output that preserves important contextual information of the extracted relational tuples (see Section 14.3.5).

Aside from assessing the merits of our discourse-aware TS approach in supporting the extraction of semantically typed relational tuples from complex input sentences in downstream Open IE applications, we evaluate the performance of our reference Open IE implementation Graphene through a comparative analysis with state-of-the-art Open IE approaches (see Section 14.3.3.1).

14.3.1 Baselines

To demonstrate the use of our discourse-aware TS approach for downstream Open IE tasks, we conduct a comparative analysis examining the output of the following state-of-the-art Open IE approaches:[9]

(i) REVERB, an early Open IE approach that is based on hand-crafted extraction patterns;

(ii) its successor system OLLIE;

(iii) ClausIE, which is the first Open IE system that is based on the idea of incorporating a sentence re-structuring stage;

(iv) Stanford Open IE, which follows a similar approach by splitting complex sentences into a set of logically entailed shorter utterances;

(v) PropS, an Open IE approach that applies a semantically-oriented sentence representation as the basis for the extraction of relational tuples;

(vi) OLLIE's successor OpenIE-4;

(vii) MinIE, which is built on top of ClausIE with the objective of minimizing both relational and argument phrases;

---

[9] We use the default configuration of each system.

(viii) OpenIE-5, which was presented as OpenIE-4's succeeding system; and
 (ix) the first end-to-end neural approach to the task of Open IE, RnnOIE.

These systems also serve as baselines for evaluating the performance of our reference
Open IE implementation Graphene, which applies a hand-crafted pattern to extract
relations and their arguments from the simplified sentences that were generated by
our proposed TS approach in the previous stage.

14.3.2 Benchmarks

To evaluate the performance of our reference Open IE implementation Graphene,
as well as to examine the use of sentence splitting as a pre-processing step for state-
of-the-art Open IE systems, we conducted experiments on three recently proposed
Open IE benchmarks:

OIE2016 The first dataset we used is Stanovsky and Dagan (2016)'s OIE2016
benchmark framework, which has become the de facto standard for the evaluation
of the task of Open IE. It consists of 10,359 extractions over 3,200 sentences from
Wikipedia and newswire articles from the Wall Street Journal.

CaRB Furthermore, we assessed and compared the quality of the output produced
by the Open IE systems under consideration on the CaRB benchmark, a crowdsourced Open IE dataset. It is comprised of a subset of 1,282 sentences from the
OIE2016 corpus, which were manually annotated using Amazon Mechanical Turk.

WiRe57 Finally, we make use of WiRe57, a small, but high-quality Open IE dataset.
It was manually curated by two Open IE experts who performed the task of Open IE
on 57 sentences taken from Wikipedia and newswire articles, resulting in a ground
truth reference of 343 relational tuples.

14.3.3 Comparative Analysis of the Outputs

To evaluate the impact of sentence splitting on the coverage and accuracy of state-of-
the-art Open IE approaches, we conduct a comparative analysis using the Open IE
benchmarks described above. In addition, we compare our reference implementation
Graphene against current Open IE systems, thus assessing the quality of its output.

14.3.3.1 Performance of the Reference Open Information Extraction Implementation Graphene

In a first step, we investigate the quality of the output generated by our reference Open IE implementation Graphene in comparison with a variety of state-of-the-art Open IE systems in an automatic comparative analysis. For this purpose, we integrate Graphene into the above-mentioned Open IE benchmarks.

In order to bring Graphene's output more into line with that of the other Open IE systems and to better fit the annotated gold relations in the benchmark datasets, we slightly modify it as follows (Cetto, 2017):

Appositive phrases Though the benchmarks aim for capturing the full set of relations asserted by a sentence, we noticed that a lot of tuples extracted from appositions, such as $\langle Barack\ Obama;\ was;\ former\ U.S.\ President \rangle$ from the phrase *"former U.S. President Barack Obama"*, are not covered in the gold relations. Hence, to filter out such extractions and thus prevent them from being marked as false positives, we use a simple heuristic: all tuples with a context level ≥ 1 and a predicate that corresponds to one of the terms $\{is,\ are,\ was,\ were\}$ are removed from Graphene's final output.

Purpose extractions Simple contexts $sc \in C_S$ that are linked to an extraction $\langle arg_1; rel\_phrase; arg_2 \rangle$ by a "Purpose" relationship are converted into separate extractions $\langle arg_{n1}; rel\_phrase_n; arg_{n2} \rangle$, where

- $rel\_phrase_n \leftarrow$ The leftmost occurrence of a verb within sc.
- $arg_{n1} \leftarrow$ The concatenation of arg_1, rel_phrase and arg_2.
- $arg_{n2} \leftarrow$ The text that follows rel_phrase_n within sc.

For instance, the context $S : \text{PURPOSE}(\textit{"to sell its German-made touring sedans in the U.S."})$ of the parent extraction $\langle Ford\ Motor\ Co.;\ created;\ the\ Merkur\ nameplate \rangle$ is transformed into the new stand-alone relational tuple $\langle Ford\ Motor\ Co.\ created\ the\ Merkur\ nameplate;\ sell;\ its\ German-made\ touring\ sedans\ in\ the\ U.S. \rangle$.

Attribution extractions Similarly, contexts $sc \in C_S$ of the type attribution that are associated with a superordinate extraction $\langle arg_1; rel\_phrase; arg_2 \rangle$ are converted into separate extractions by constructing a new relational tuple of the form $\langle arg_{n1}; rel\_phrase_n; arg_{n2} \rangle$, where

- $rel\_phrase_n \leftarrow$ The leftmost occurrence of a verb within c.
- $arg_{n1} \leftarrow$ The text that precedes rel_phrase_n within c.
- $arg_{n2} \leftarrow$ The concatenation of arg_1, rel_phrase and arg_2.

For example, when given the extraction ⟨*previous studies of flashbulb memories; are limited; by the reliance on small sample groups*⟩ together with the context S:ATTRIBUTION(*"some researchers recognized"*), the latter is converted into the following relational tuple: ⟨*some researchers; recognized; previous studies of flashbulb memories are limited by the reliance on small sample groups*⟩.

N-ary relations By default, Graphene extracts only binary relations. However, the gold standard benchmark datasets we use are composed of n-ary relations. To adapt Graphene's output to this feature, we add all simple contexts $sc \in C_S$ of an extraction as additional arguments besides arg_1 and arg_2. For instance, consider the binary extraction ⟨*Mount Apo; was declared; a national park*⟩ that is associated with the following three simple contexts: *"on May 9, 1936"*, *"with Proclamation no. 59"* and *"by President Manuel L. Quezon"*. In order to turn this output into an n-ary relation, the simple contexts are appended to their parent tuple as arg_3, arg_4 and arg_5, resulting in the following 5-ary extraction: ⟨*Mount Apo; was declared; a national park; on May 9, 1936; with Proclamation no. 59; by President Manuel L. Quezon*⟩.

Graphene and the Open IE baseline systems are evaluated according to the following metrics:

- We compute the PR curve by evaluating the systems' performance at different extraction confidence thresholds.
- We calculate the AUC as a scalar measure of the systems' overall performance.
- We report the average precision and recall of the extracted relational tuples, as well as their F_1 score.

Currently, Graphene does not assign confidence scores to its extractions. Hence, to allow for assessing Graphene's performance at different confidence thresholds, we determined a simple mathematical function that calculates the confidence score $conf_s$ of an extraction based on the length n of the source sentence s (Cetto, 2017):

$$conf_s(n) \leftarrow \frac{1}{0.005n + 1} \tag{14.1}$$

This function is based on the assumption that the longer the input, the more difficult and thus error-prone is the extraction of relations and their corresponding arguments. For instance, it yields a confidence level of 80% for a sentence with 50 characters, while a sentence containing 200 characters results in a confidence score of only 50%.

To compare the performance of Graphene to state-of-the-art Open IE systems, we integrated it into the following benchmarks:

OIE2016   To allow for a comparison with current approaches, we extended the OIE2016 benchmark by incorporating more recent Open IE systems, including MinIE, OpenIE-5 and RnnOIE.[10] Moreover, since MinIE and PropS do not provide confidence scores for their extractions, we apply Graphene's confidence function from Equation 14.1.[11]

To match predicted extractions to reference extractions, the OIE2016 benchmark serializes both the predicted and the gold tuples into a sentence and computes their lexical match. A pair of predicted-reference tuples is declared a match if the fraction of words of the prediction that is also present in the reference surpasses a pre-defined threshold (0.25 by default).[12] Based on this lexical matching function, a set of gold tuples is compared with a set of predicted tuples to estimate word-level precision and recall, from which the systems' AUC score and PR curve are derived.

However, the scoring method used in the OIE2016 benchmark for matching predictions to ground truth tuples has been criticized for not being robust (Bhardwaj et al., 2019; Lechelle et al., 2019). One of its major weaknesses is that the matching function does not penalize extractions that are overly long. Since it only checks whether a sufficient number of tokens from the reference occurs in the predicted spans, extractions with long relational or argument phrases have a natural advantage over shorter extractions, as they are more likely to contain more words from the reference tuples. Moreover, the scoring function does not penalize extractions for misidentifying parts of a relation in an argument slot (or vice versa), since it concatenates the relational phrase and its arguments to a continuous text span. Finally, the scorer loops over gold tuples in an arbitrary order and matches them to predicted extractions in a sequential manner; once a gold matches to a predicted extraction, it is rendered unavailable for any subsequent, potentially better fitting extraction.

---

[10] We used the latest version of the OIE2016 benchmark (commit `6d9885e50c7c02cbb41ee98-ccce9757fda94c89c` on their github repository).

[11] Though the authors of ClausIE specify a method for calculating the confidence scores of their system's extractions, it is not implemented in the published source code. Therefore, we make use of Graphene's confidence score function for ClausIE's extractions, too.

[12] The original scoring function presented in Stanovsky and Dagan (2016) was more restrictive. Here, a prediction was classified as being correct if the predicted and the reference tuple agree on the grammatical head of all their elements, i.e. the predicate and the arguments. However, this scheme was later relaxed in their github repository to the lexical matching function described above, which has become the OIE2016 benchmark's default scorer in the current literature.

CaRB To overcome the limitations of the scoring function used in the OIE2016 benchmark, Bhardwaj et al. (2019) presented an improved matching function addressing the issues described above. Instead of a lexical match, CaRB's scorer utilizes a tuple match, i.e. it matches relational phrase with relational phrase and argument slot with argument slot. Moreover, CaRB creates an all-pair matching table, with each column representing a predicted tuple and each row representing a gold tuple, rather than greedily matching references to predictions in arbitrary order. Between each pair of tuples, precision and recall are computed. For determining the overall recall, the maximum recall score is taken in each row, and averaged, while for computing the overall precision, each system's predictions are matched one to one with gold tuples, in the order of best match score to worst; the precision scores are then averaged to get the overall precision score. To calculate the PR curve, this computation is done at different confidence thresholds of the predicted extractions. Note that CaRB uses a multi-match approach for recall computation, which prevents from overly penalizing systems that merge information from multiple gold tuples into a single extraction. In contrast, its precision computation is based on a single match approach between predictions and references, penalizing systems that produce multiple very similar and redundant extractions.

For the evaluation of the Open IE approaches, we do not use the CaRB gold tuples because they are ill-suited for our purposes for a variety of reasons. First, though atomicity was one of the declared goals when creating the ground truth reference, we noticed that CaRB's gold tuples contain a lot of overgenerated predicates and overly long argument phrases. This is contrary to our intentions, as one of the central goals of our approach is to generate extractions where each tuple represents an indivisible unit that cannot be further decomposed into meaningful propositions. Moreover, CaRB's reference tuples often convey inferred relations that are not explicitly stated in the input sentences. Conversely, our Open IE approach focuses on the extraction of relations that are explicitly expressed in the text, rather than extracting information that is merely implied by it. Finally, the annotators of the CaRB corpus only marked up binary relations (optionally augmented by temporal and local information), while our approach aims to capture the full context of a relation. Therefore, we chose to keep the ground truth relations of the OIE2016 corpus, and only use the improved evaluation metrics of the CaRB benchmark.[13]

Similar to the OIE2016 benchmark, we integrated MinIE and RnnOIE into the CaRB framework to allow for a comparison with more recent approaches.

---

[13] The results of the comparative analysis on the original CaRB dataset can be found in Section 22 in the online supplemental material.

WiRe57 To examine the transferability of the results to other datasets,[14] we included the WiRe57 corpus in our analysis. Ignoring the confidence values of the predicted extractions, it is limited to the computation of precision, recall and F_1, which are measured at the level of tokens. Precision is computed as the proportion of the words in the prediction that are found in the reference tuple, while recall represents the proportion of reference words found in the predicted tuple. To match predictions to gold tuples, the pair with the maximum F_1 score is greedily removed from the pool, until no remaining tuples match. The systems' overall performance is the token-weighted precision and recall over all tuples. As before, with OpenIE-5 and RnnOIE, we incorporated more recent approaches to the task of Open IE into the benchmark framework.

The analysis described above allows for assessing the performance of our reference implementation Graphene as compared to state-of-the-art Open IE approaches in terms of coverage and accuracy. In addition, we calculate a variety of descriptive statistics to get a better understanding of the similarities and differences of the output produced by the Open IE systems under consideration:

   (i)   the total number of relational tuples extracted from the input (#E);

   (ii) the average number of facts extracted per source sentence (#E/S);

  (iii) the percentage of non-redundant extractions, i.e. extractions that do not contain subsequences of other tuples (#$E_{non\text{-}redundant}$);

  (iv) the number of sentences from which not even one fact is extracted (#S < 1E);

   (v) the average length of the extracted relational tuples in total (in terms of the average number of tokens), as well as the average length of the phrases in subject, predicate and object position each;

  (vi) the percentage of tokens from the input that are contained in at least one of the extracted relational tuples, serving as a proxy for the completeness of the extractions (coverage);

 (vii) distribution of the arity of the extracted relations over the dataset.

14.3.3.2 Sentence Splitting as a Pre-processing Step

We hypothesize that the complexity of the relation extraction task in Open IE systems will be reduced by taking advantage of the split sentences instead of dealing with the raw complex source sentences. This assumption is based on the idea that shorter sentences with a more regular structure are easier to process and analyze for downstream Open IE applications. Hence, we expect to improve the performance of state-of-the-art Open IE systems in terms of the precision and recall of the

---

[14] CaRB is built over a subset of OIE2016's original sentences.

extracted relational tuples when operating on the simplified sentences generated by our discourse-aware TS approach. To examine this hypothesis, we compare the performance of the Open IE approaches listed in Section 14.3.1 when directly operating on the raw input data with their performance when our reference TS implementation DISSIM is used as a pre-processing step to split complex source sentences into a set of syntactically simplified sentences with a more regular structure.

For this experiment, we use the same setup as before in Section 14.3.3.1. However, we restrict our study to the OIE2016 benchmark and the customized version of the CaRB framework, leaving aside the WiRe57 dataset due to its limited informative value because of its very small size. In two consecutive rounds, each Open IE system is again quantitatively evaluated by calculating the PR curve of its extractions at different confidence thresholds, by computing the AUC score and by measuring average precision, recall and F_1 of the extracted relational tuples. Analogous to the experiment from Section 14.3.3.1, in the first round, each system is executed on the original complex input sentences, whereas in the second round, the simplified sentences generated by our discourse-aware TS approach are taken as input. The setup of the second round corresponds to the setting we used for evaluating the output of our reference Open IE implementation Graphene in the previous section, i.e. some minor modifications are carried out to remove extractions from appositive phrases, and to transform contexts with a purpose or attribution relationship into separate extractions. Furthermore, the remaining types of simple contextual arguments are attached as additional arguments to their parent extraction to generate n-ary relations. In that way, we obtain a representation that better reflects the benchmarks' gold standard relational tuples. Based on the results we obtain we can make a statement about how sentence splitting affects the performance of state-of-the-art Open IE systems in terms of *coverage* and *accuracy* of the extracted relations when applied as a pre-processing step.

On top of that, we calculate the same set of descriptive statistics as in the previous section, allowing us to determine how the characteristics of the extracted relational tuples change when the tested Open IE systems operate on the simplified sentences instead of processing the original complex input data.

14.3.4 Manual Analyses

To complement the automatic evaluation from the previous section, which compared the output of our reference Open IE implementation Graphene with the above-mentioned baseline systems and assessed the impact of sentence splitting on their performance by means of a set of automatic metrics, we conduct a manual analysis for a more in-depth examination of the extracted relational tuples.

14.3.4.1 Error Analysis

In order to get further insights into the quality of the output produced by our reference Open IE implementation Graphene, we performed a detailed error analysis. For this purpose, we first randomly sampled a set of 200 sentences from the OIE2016 corpus. When executing our Open IE approach Graphene on them, 414 relations were extracted. We then manually analyzed the generated relational tuples, distinguishing correct extractions from inaccurate ones. In a second step, we grouped the incorrect tuples into different classes, figuring out the major types of errors that can be observed in the relations extracted by Graphene. Thus, the results of this investigation may suggest starting points for prospectively improving our approach.

14.3.4.2 Qualitative Analysis of the Extracted Relations

Furthermore, we carried out a qualitative analysis of the outputs produced by Graphene and the baseline Open IE systems listed in Section 14.3.1, where we compare and discuss the peculiarities of the relations extracted by each approach based on a representative example. In a second step, we extend this analysis to examine the effects on the generated output when operating on the semantic hierarchy of minimal propositions instead of dealing with the original source sentences.

14.3.5   Analysis of the Lightweight Semantic Representation of Relational Tuples

Finally, we investigated whether the semantic hierarchy of minimal propositions generated by our discourse-aware TS approach can be leveraged to transform the shallow semantic representation of state-of-the-art Open IE systems into a canonical context-preserving representation of relational tuples. For this purpose, we analysed the core characteristics of the resulting representation in the light of the experiments outlined above. The aim was to determine to what extent the semantic hierarchy of simplified sentences supports

(a) the *extraction of simplistic canonical predicate-argument structures* from complex source sentences, and

(b) the *enrichment of the extracted relational tuples with additional meta information* to produce an output that preserves important contextual information.

14.4   Sentence Splitting Corpus

To assess the quality of MINWIKISPLIT, the sentence splitting corpus we compiled by leveraging our proposed syntactic TS approach, we performed both a manual analysis and an automatic evaluation of the split sentences.

14.4.1 Automatic Metrics

Analogous to Section 14.1.3.1, we computed a number of descriptive statistics on a random sample of 1,000 complex-simple sentence pairs to estimate the quality of the simplified target sentences of the MINWIKISPLIT corpus, including

(i)   the average sentence length of the simplified sentences in terms of the average number of tokens per output sentence (#T/S);

(ii) the average number of simplified output sentences per complex input (#S/C);

(iii) the percentage of sentences that are copied from the source without performing any simplification operation (%SAME), serving as an indicator for conservatism, i.e. the tendency to retain the input rather than transforming it; and

(iv) the averaged word-based Levenshtein distance from the input (LD_{SC}), which provides further evidence for how reluctant the underlying system is in splitting the input into simplified sentences.

Moreover, to measure the structural simplicity of the instances contained in MINWIKISPLIT, we calculated the SAMSA and $SAMSA_{abl}$ scores of both the complex source and the simplified output sentences (see Section 14.1.3.3).

14.4.2 Manual Analysis

In a second step, we randomly selected a subset of 300 sentences from MINWIKISPLIT, on which we conducted a manual analysis in order to get some deeper insights into the quality of the simplified sentences. Following the approach detailed in Section 14.1.4.1, each pair of complex source and simplified target sentences was rated by two non-native, but fluent English speakers according to the three parameters of grammaticality, meaning preservation and structural simplicity on a 5-point Likert scale.

Results and Discussion   15

Below, we present and discuss the results of our experiments. We evaluated the performance of our proposed discourse-aware TS approach in three respects:

(1) To assess its performance with regard to the sentence splitting subtask, we compared the output generated by our reference implementation DISSIM against several state-of-the-art syntactic TS baselines, based on a set of automatic metrics and manual analyses (see Section 15.1).

(2) In a second step, we determined the accuracy of the semantic hierarchy that is established between the split sentences by our TS framework, again on the basis of both an automatic and a human evaluation (see Section 15.2).

(3) Finally, we examined the merits of our discourse-aware TS approach in supporting the extraction of semantically typed relational tuples from complex input sentences in downstream Open IE applications (see Section 15.3).

In addition, we investigated the quality of the simplified target sentences in the sentence splitting corpus we compiled by leveraging our proposed TS approach (see Section 15.4).

Supplementary Information The online version contains supplementary material available at https://doi.org/10.1007/978-3-658-38697-9_15.

15.1   Subtask 1: Splitting into Minimal Propositions

In the following, we describe the evaluation results with respect to the first subtask of our approach, whose goal is to decompose sentences that present a complex linguistic structure into a set of syntactically simplified propositions, with each of them representing a minimal semantic unit.

15.1.1 Automatic Metrics

To determine how good the simplification output produced by our discourse-aware TS framework is, we first calculated a number of automatic evaluation metrics.

15.1.1.1 WikiLarge Corpus

Descriptive Statistics Table 15.1 reports the results that were achieved on the 359 test sentences from the WikiLarge corpus, based on a set of descriptive statistics that allow for getting a first impression of the simplified output. Transforming each input sentence of the dataset, our reference implementation DISSIM reaches the highest splitting rate among the TS systems under consideration, together with HYBRID,

Table 15.1 Automatic evaluation results on the WikiLarge corpus. We report evaluation scores for a number of descriptive statistics. These include the average sentence length of the simplified sentences in terms of the average number of tokens per output sentence (#T/S) and the average number of simplified output sentences per complex input (#S/C). Furthermore, in order to assess system conservatism, we measure the percentage of sentences that are copied from the source without performing any simplification operation (%SAME) and the averaged Levenshtein distance from the input (LD_{SC}). The highest score by each evaluation criterion is shown in bold. (*) from Zhang and Lapata (2017)

	#T/S	#S/C	% SAME	LD_{SC}
Complex source	22.06	1.03	100.00	0.00
Simple reference*	20.19	1.14	0.00	7.14
DISSIM	**11.01**	**2.82**	**0.00**	11.90
DSS	12.91	1.87	**0.00**	8.14
SENTS	14.17	1.09	**0.00**	**13.79**
HYBRID	13.44	1.03	**0.00**	13.04
YATS	18.83	1.40	18.66	4.44
RegenT	18.20	1.45	41.50	3.77

DSS and SENTS. With 2.82 split sentences per input on average, our framework outputs by a large margin the highest number of structurally simplified sentences per source. Moreover, consisting of 11.01 tokens on average, the DISSIM approach returns the shortest sentences of all systems, on an average halving the length of the input. The relatively high word-based Levenshtein distance of 11.90 confirms previous findings, suggesting that our structural TS approach tends to extensively split the source sentences instead of retaining the input by simply copying the input without performing any simplification operations. SENTS (13.79) is the only system that shows an even higher Levenshtein distance to the source.

Translation Assessment With regard to SARI, our DISSIM framework (35.05) again outperforms the baseline systems on WikiLarge, as illustrated in Table 15.2. However, according to the findings presented in Sulem et al. (2018a), the significance of this metric for evaluating TS systems that perform sentence splitting operations is very limited. The authors demonstrate that SARI does not correlate with human judgments on structural simplicity, which might be due to its focus on lexical rather than syntactic TS rewrites (see Section 21). Moreover, their study indicates that SARI shows a negative correlation with human ratings on grammaticality and meaning preservation. Though, this relationship vanishes when restricting the analysis to sentence-level correlations on a dataset that explicitly targets the sentence splitting task.

In addition, we compared the output generated by our TS framework against the baseline systems using the BLEU metric. Here, our approach is among the

Table 15.2 Automatic evaluation results on the WikiLarge corpus. We calculate the average BLEU and SARI scores for the rephrasings of each system. The highest score by each evaluation criterion is shown in bold. (*) from Zhang and Lapata (2017)

	BLEU	SARI
Complex source	91.25	32.53
Simple reference*	99.48	43.09
DISSIM	63.03	**35.05**
DSS	74.42	34.32
SENTS	54.37	29.76
HYBRID	48.97	26.19
YATS	73.07	33.03
RegenT	**82.49**	32.41

systems with the lowest score (63.03) on the WikiLarge dataset. However, Sulem et al. (2018a) demonstrated that BLEU, too, is ill suited for the evaluation of TS approaches that perform sentence splitting transformations. Their study reveals that in this case, BLEU negatively correlates with structural simplicity, thus penalizing sentences that present a simplified syntax. Moreover, the authors show that this metric presents no correlation with human judgments on grammaticality and meaning preservation. Hence, the results based on the SARI and BLEU metrics must be treated with caution. We only report these scores for the sake of completeness and to match past work.

Table 15.3 Automatic evaluation results on the WikiLarge corpus. We calculate the SAMSA and SAMSA$_{abl}$ scores of the simplified output, which are the first metrics targeting syntactic aspects of TS. The highest score by each evaluation criterion is shown in bold. (*) from Zhang and Lapata (2017)

	SAMSA	SAMSA$_{abl}$
Complex source	0.59	0.96
Simple reference*	0.48	0.78
DISSIM	**0.67**	0.84
DSS	0.64	0.75
SENTS	0.40	0.58
HYBRID	0.47	0.76
YATS	0.56	0.80
RegenT	0.61	**0.85**

Syntactic Complexity According to Sulem et al. (2018b), the SAMSA and SAMSA$_{abl}$ metrics are better suited for the evaluation of the sentence splitting task. With a score of 0.67 on the WikiLarge corpus, the DISSIM framework shows the best performance for SAMSA, while its score of 0.84 for SAMSA$_{abl}$ is just below the one obtained by the RegenT system (0.85) (see Table 15.3). As reported in Sulem et al. (2018b), SAMSA highly correlates with human judgments for structural simplicity and grammaticality, while SAMSA$_{abl}$ achieves the highest correlation for meaning preservation.

15.1.1.2 Newsela Corpus

The results on the Newsela dataset, depicted in Table 15.4, support our findings from the WikiLarge corpus, indicating that our TS approach can be applied in a domain independent manner.

Table 15.4 Automatic evaluation results on the 1,077 test sentences from the Newsela corpus. The highest score by each evaluation criterion is shown in bold

	#T/S	#S/C	%SAME	LD$_{SC}$	BLEU	SARI	SAMSA	SAM-SA$_{abl}$
Complex	23.34	1.01	100.00	0.00	20.91	9.84	0.49	0.96
Simple	12.81	1.01	0.00	16.25	100	91.13	0.25	0.46
DisSim	**11.20**	**2.96**	**0.00**	13.00	14.54	**49.00**	**0.57**	0.84
Hybrid	12.49	1.02	**0.00**	**13.46**	14.42	40.34	0.38	0.74
YATS	18.71	1.42	16.16	5.03	17.51	36.88	0.50	0.83
RegenT	16.74	1.61	33.33	5.03	**18.96**	32.83	0.55	**0.85**

15.1.1.3 WikiSplit Corpus

Table 15.5 illustrates the scores achieved for the automatic evaluation metrics on the WikiSplit dataset. Though the Copy512 system beats our approach in terms of BLEU and SARI, the remaining scores are clearly in favour of the DisSim system. With an average length of 11.91 tokens per simplified sentence, it reduces the source sentences to about one third of their original length. The tendency to perform a large amount of splitting operations is underlined by both the high number of simple sentences per complex input (4.09) and the large Levenshtein distance from the source (19.10), as well as the low percentage of unmodified input sentences (0.76%). The SAMSA (0.54) and SAMSA$_{abl}$ scores (0.84) convey further evidence that our approach succeeds in producing structurally simplified sentences that are grammatically sound and preserve the original meaning of the input.

Table 15.5 Automatic evaluation results on the 5,000 test sentences from the WikiSplit dataset. The highest score by each evaluation criterion is shown in bold

	#T/S	#S/C	% SAME	LD$_{SC}$	BLEU	SARI	SAMSA	SAM-SA$_{abl}$
Complex	32.01	1.10	100.00	0.00	74.28	29.91	0.37	0.95
Simple	18.14	2.08	0.00	7.48	100	94.71	0.49	0.75
DisSim	**11.91**	**4.09**	**0.76**	**19.10**	51.96	39.33	**0.54**	**0.84**
Copy512	16.55	2.08	13.30	2.39	**76.42**	**61.51**	0.51	0.78

15.1.1.4 Rule Application Statistics

We calculated the frequency distribution of the grammar rules that were triggered by our reference implementation DISSIM during the transformation of the sentences from WikiLarge, Newsela and WikiSplit. The results are displayed in Table 15.6b for clausal disembedding and Table 15.6a for the decomposition of phrasal elements. The ten most matched rules are shown in bold for each dataset. It becomes apparent that the transformation patterns that are applied the most often are the same for all

Table 15.6 Rule application statistics

CLAUSAL TYPE	RULE	WIKI-LARGE	NEWSELA	WIKISPLIT	TOTAL
Coordinate clauses	RULE #1	**38**	**123**	**1388**	**1549 (8.77%)**
Adverbial clauses	RULE #2	**40**	**123**	**832**	**995 (5.63%)**
	RULE #3	6	59	189	254 (1.44%)
	RULE #4	10	49	221	280 (1.59%)
	RULE #5	0	6	8	14 (0.08%)
	RULE #6	0	0	3	3 (0.02%)
	RULE #7	0	0	2	2 (0.01%)
Relative clauses (non-restrictive)	RULE #8	1	2	33	36 (0.20%)
	RULE #9	6	16	127	149 (0.84%)
	RULE #10	1	0	6	7 (0.04%)
	RULE #11	0	0	15	15 (0.08%)
	RULE #12	10	55	377	442 (2.50%)
Relative clauses (restrictive)	RULE # 13	0	0	0	0 (0.00%)
	RULE #14	0	0	17	17 (0.10%)
	RULE #15	**24**	**121**	**612**	**757 (4.29%)**
	RULE #16	7	35	254	296 (1.68%)
Reported speech	RULE #17	2	33	38	73 (0.41%)
	RULE #18	0	4	0	4 (0.02%)
	RULE #19	0	22	0	22 (0.12%)
	RULE #20	3	**99**	140	242 (1.37%)

(a) Clausal disembedding

(continued)

Table 15.6 (continued)

Phrasal Type	Rule	Wiki-Large	Newsela	WikiSplit	Total
Coordinate verb phrases	Rule #21	38	107	1268	1413 (8.00%)
Coordinate noun phrases	Rule #22	51	81	959	1091 (6.18%)
	Rule #23	7	32	168	207 (1.17%)
Participial phrases	Rule #24	8	11	266	285 (1.61%)
	Rule #25	40	76	652	768 (4.35%)
	Rule #26	23	38	405	466 (2.64%)
	Rule #27	4	7	156	167 (0.95%)
Appositions (non-restrictive)	Rule #28	38	147	844	1029 (5.83%)
Appositions (restrictive)	Rule #29	32	66	863	961 (5.44%)
Prepositional phrases	Rule #30	117	421	2965	3503 (19.83%)
	Rule #31	65	152	1314	1531 (8.67%)
	Rule #32	21	66	490	570 (3.23%)
Adjectival/adverbial phrases	Rule #33	17	46	194	257 (1.46%)
	Rule #34	14	22	212	248 (1.40%)
Lead noun phrases	Rule #35	0	0	2	2 (0.01%)
Total		623	2019	15020	17662

(b) Phrasal disembedding

three corpora, with the notable exception of Rule #20, which is among the top ten rules for Newsela, but not for the other two datasets. This pattern encodes a form of reported speech. Since this type of speech is far more prevalent in newswire texts than in Wikipedia articles, this represents a result that was to be expected.

Moreover, the frequency distribution reveals that there is roughly one rule per group of linguistic constructs that is among the top ten of the most fired transformation patterns. Furthermore, Tables 15.6a and 15.6b show that the ten most matched simplification rules (i.e., less than 30% of the 35 manually defined grammar rules) account for about 75% of the rule applications. Hence, we conclude that the specified transformation patterns cover the majority of syntactic phenomena that are involved in complex sentence constructions. In combination with the findings of the human evaluation in Section 15.1.2.1 and the figures computed in Section 15.1.1, revealing that our reference implementation DISSIM generates the shortest output sentences on average, while considerably reducing their syntactic complexity and preserving most of the information contained in the input, we deduce that our proposed structural TS approach largely succeeds in splitting complex source sentences into a sequence of atomic semantic units that present a simplified syntax.

15.1.2 Manual Analyses

In the following, we report the results of the human ratings on the quality of the simplified output and the manual qualitative analysis of the linguistically principled transformation patterns.

15.1.2.1 Human Judgments

The results of the human evaluation are displayed in Table 15.7. The inter-annotator agreement was calculated using Cohen's quadratic weighted κ Cohen (1968), resulting in rates of 0.72 (G), 0.74 (M) and 0.60 (S). Hence, the figures indicate moderate to substantial agreement between the annotators, suggesting that the scores we got for the three categories under consideration present a reliable result. The assigned scores demonstrate that our reference implementation DISSIM outperforms all other TS systems in the simplicity dimension (S). With a score of 1.30 on the WikiLarge sample sentences, it is far ahead of the baseline approaches, with HYBRID (0.86) coming closest. However, this system receives the lowest scores for grammaticality (G) and meaning preservation (M). RegenT obtains the highest score for G (4.64), while YATS is the best-performing approach in terms of M (4.60). With a rate of only 0.22, though, it achieves a low score for S, indicating that the high score in the M dimension is due to the conservative approach taken by YATS, resulting in only a small number of simplification operations. This explanation also holds true for RegenT's high mark for G. Still, our DISSIM approach follows closely, with a score of 4.50 for M and 4.36 for G, suggesting that it obtains its goal of returning fine-grained simplified sentences that achieve a high level of grammaticality and preserve the meaning of the input. Considering the average scores of all systems under consideration, our approach is the best-performing system (3.39), followed

Table 15.7 Human evaluation ratings on a random sample of 50 sentences per dataset. Grammaticality (G) and meaning preservation (M) are measured using a 1 to 5 scale. A −2 to 2 scale is used for scoring the *structural* simplicity (S) of the output relative to the input sentence. The last column *(average)* presents the average score obtained by each system with regard to all three dimensions. The highest score by each evaluation criterion is shown in bold

	G	M	S	average
Simple reference	4.70	4.56	−0.2	3.02
DISSIM	4.36	4.50	**1.30**	**3.39**
DSS	3.44	3.68	0.06	2.39
SENTS	3.48	2.70	−0.18	2.00
HYBRID	3.16	2.60	0.86	2.21
YATS	4.40	**4.60**	0.22	3.07
RegenT	**4.64**	4.56	0.28	3.16

(a) WikiLarge test set

	G	M	S	average
Simple reference	4.92	2.94	0.46	2.77
DISSIM	4.44	4.60	**1.38**	**3.47**
HYBRID	2.97	2.35	0.93	2.08
YATS	4.26	4.42	0.32	3.00
RegenT	**4.54**	**4.70**	0.62	3.29

(b) Newsela test set

	G	M	S	average
Simple reference	4.72	4.32	0.44	3.16
DISSIM	4.36	4.36	**1.66**	**3.46**
Copy512	**4.72**	**4.72**	0.92	3.45

(c) WikiSplit test set

by RegenT (3.16). The human evaluation ratings on the Newsela and WikiSplit sentences show similar results, again supporting the domain independence of our proposed approach.

15.1.2.2 Qualitative Analysis of the Transformation Patterns

Coverage and Accuracy of the Transformation Rules Table 15.8 shows the results of the manual qualitative analysis of the 35 linguistically principled transformation patterns. With an average rate of more than 80% for the syntactic phenomena under consideration, the overall ratio of the simplification rules being fired whenever the corresponding clausal or phrasal construct is present in the given input sentence is very high. However, the patterns for the extraction of coordinate noun phrases, as well as prepositional phrases exhibit a much lower recall ratio. Regarding the former, the main reason is that we follow a very conservative approach when splitting lists of noun phrases, since this type of syntactic construct is prone to parsing errors

Table 15.8 Qualitative analysis of the transformation rule patterns. This table presents the results of a manual analysis of the performance of the hand-crafted simplification patterns. The first column lists the syntactic phenomenon under consideration, the second column indicates its frequency in the dataset, the third column displays the percentage of the grammar fired, and the fourth column reveals the percentage of sentences where the transformation operation results in a correct split

	frequency	%fired	%correct transformations
Clausal disembedding			
Coordinate clauses	113	93.8%	99.1%
Adverbial clauses	113	84.1%	96.8%
Relative clauses (non-restrictive)	108	88.9%	70.8%
Relative clauses (restrictive)	103	86.4%	75.3%
Reported speech	112	82.1%	75.0%
Phrasal disembedding			
Coordinate verb phrases	109	85.3%	89.2%
Coordinate noun phrases	115	48.7%	82.1%
Participial phrases	111	76.6%	72.9%
Appositions (non-restrictive)	107	86.0%	83.7%
Appositions (restrictive)	122	87.7%	72.0%
Prepositional phrases	163	68.1%	75.7%
Total	1276	80.7%	81.1%

and hard to distinguish from appositions. When decomposing prepositional phrases, we are confronted with the complex problem of resolving attachment ambiguities that is pervasive when dealing with this kind of phrasal element Bailey et al. (2015), Gelbukh and Calvo (2018). Here, too, we chose a rather conservative approach, restricting the extraction of prepositional phrases to those that are either offset by commas or belong to a certain subclass acting as complements of verb phrases.

With respect to the accuracy of the specified transformation patterns in terms of the percentage of input sentences that were correctly split into syntactically simplified output sentences, the rules for disembedding coordinate and adverbial clauses perform remarkably well, approaching an accuracy rate of almost 100%. On average, correct transformations are carried out in over 80% of the cases. The syntactic construct that provides the lowest accuracy are non-restrictive relative clauses, which are prone to missing some essential part from the source sentence or assigning the wrong attachment phrase to the extracted clausal component, as revealed by the error analysis.

Table 15.9 Results of the error analysis. Six types of errors were identified (Error 1: additional parts; Error 2: missing parts; Error 3: morphological errors; Error 4: wrong split point; Error 5: wrong referent; Error 6: wrong order of the syntactic elements)

	Err. 1	Err. 2	Err. 3	Err. 4	Err. 5	Err. 6
Clausal disembedding						
Coordinate clauses	1	0	0	0	0	0
Adverbial clauses	1	1	0	1	0	0
Relative clauses (non-restr.)	5	8	0	0	14	1
Relative clauses (restrictive)	8	8	2	0	5	1
Reported speech	5	1	13	1	2	1
Phrasal disembedding						
Coordinate verb phrases	4	3	2	1	0	0
Coordinate noun phrases	3	3	0	3	1	0
Participial phrases	2	2	4	2	13	0
Appositions (non-restrictive)	0	5	3	0	7	0
Appositions (restrictive)	1	21	3	0	0	0
Prepositional phrases	3	11	4	6	4	0
Total	33	63	31	14	46	3
	(17%)	(33%)	(16%)	(7%)	(24%)	(2%)

Error analysis Representing about 60% of the erroneous simplifications, missing elements of the input when constructing the simplified sentences (error 2) and allocating the wrong referent (error 5) are the most frequently experienced problems that were identified in the qualitative analysis of the performance of the hand-crafted simplification rules. In total, six types of errors were identified, as detailed in Table 15.9. With the help of an example sentence, each error class is illustrated in Table 15.10.

Table 15.10 Error classification

TYPE	ERROR	INPUT	OUTPUT
1	additional parts	Tyler comforted Julia in her grief and won her consent to a secret engagement.	Tyler comforted Julia in her grief. Tyler **comforted Julia in her** won her consent.
2	missing parts	The magazine began publication ... **under the name** For Him.	The magazine began publication For Him.
3	morphological errors	Mars has two Moons, Phobos and Deimos.	Mars has two Moons. Phobos and Deimos **is** two moons.
4	wrong split point	Peter Ermakov killed her **with a gun shot to the left side of her head.**	Peter Ermakov killed her. **This was to the left side of her head. This was with a gun shot.**
5	wrong referent	Pushkin and **his wife Natalya Goncharova, whom he married in 1831,** later became regulars of the court society.	Pushkin and his wife Natalya Goncharova later became regulars of the court society. He married **Pushkin and his wife Natalya Goncharova** in 1831.
6	wrong order of the syntactic elements	She is the adoptive mother of actor Dylan McDermott, whom she adopted when he was 18.	She is the adoptive mother of actor Dylan McDermott. **Dylan McDermott she** adopted when he was 18.

To give a clearer picture of the source of the errors, we complemented the analysis described above with a tracing where the errors start. We found that they can be attributed to two causes: either the parse tree constructed by the constituency parser is erroneous or the simplification rules we specified fail to capture slight nuances or variations of the input, presenting a rarely occurring sentence structure. This is illustrated in the following examples:

- *"He is a southpaw boxer, who currently trains under Billy Hussein, brother of boxers Nedal and Hussein Hussein."*: Here, the appositive *"boxers"* is extracted and transformed into the simplified output *"Nedal is boxers."*, failing to convert the number of the term *"boxers"* from plural into singular.
- *"Chun was later pardoned by President Kim Young-sam on the advice of then President-elect Kim Dae-jung."*: Here, the adverb modifier *"then"* is missed when decomposing the appositive phrase *"then President-elect"*.
- *"Between 1898 and 1901 he was a choral coach and subsequently an assistant conductor at the Bayreuth Festival."*: When simplifying this sentence, the temporal phrase *"Between 1898 and 1901"* is linked to both simplified coordinate verb phrases, *"He was a choral coach."* and *"He was an assistant conductor at the Bayreuth Festival."* However, it only refers to the former, while the latter is described by the adverb of time *"subsequently"*. Thus, the meaning of the source sentence is slightly altered in the simplified output.

Table 15.11 Total number of errors that can be attributed to wrong constituency parses *(first number)* compared to rules not covering the respective sentence's structure *(second number)* (Error 1: additional parts; Error 2: missing parts; Error 3: morphological errors; Error 4: wrong split point; Error 5: wrong referent; Error 6: wrong order of the syntactic elements). The last row and column *("% erroneous parses")* indicate the percentage of errors that can be attributed to erroneous constituency parse trees

	Err. 1	Err. 2	Err. 3	Err. 4	Err. 5	Err. 6	% err. parses
Clausal disembedding							
Coordinate clauses	0–1	–	–	–	–	–	0.00
Adverbial clauses	0–1	0–1	–	1–0	–	–	0.33
Relative clauses (non-restr.)	4–1	1–7	–	–	8–6	0–1	0.46
Relative clauses (restrictive)	8–0	8–0	1–1	–	5–0	0–1	0.92
Reported speech	2–3	0–1	0–13	1–0	2–0	1–0	0.26
Phrasal disembedding							
Coordinate verb phrases	3–1	0–3	0–2	1–0	–	–	0.40
Coordinate noun phrases	3–0	0–3	–	2–1	0–1	–	0.50
Participial phrases	0–2	0–2	4–0	0–2	9–4	–	0.57
Appositions (non-restrictive)	–	5–0	1–2	–	5–2	–	0.73
Appositions (restrictive)	0–1	7–14	0–3	–	–	–	0.28
Prepositional phrases	0–3	7–4	0–4	4–2	2–2	–	0.46
% erroneous parses	0.61	0.44	0.19	0.64	0.67	0.33	0.50

As Table 15.11 shows, additional parts that are included by mistake in the simplified sentences, wrong split points and wrong referents can typically be traced
back to erroneous parses, while morphological errors and wrong orderings of the
syntactic elements in the simplified output mostly appear due to particular structures that are not covered by the specified transformation patterns. Errors that can be
attributed to erroneous parse trees will diminish as the performance of the underlying
parser increases. Thus, more attention should be paid to errors caused by underspecified simplification rules. According to Table 15.11, there is the most potential for
improvement in this regard with respect to reported speech and restrictive appositive
phrases.

15.2   Subtask 2: Establishing a Semantic Hierarchy

In the following section, we describe the results of both the automatic evaluation
and the manual analysis with regard to the second subtask, whose goal is to set up
a semantic hierarchy between the split sentences.

15.2.1 Automatic Metrics

Using the matching function described in Section 14.2.1, we obtained 1,827 matched
sentence pairs, i.e. 11.74% of the pairs of simplified sentences were successfully
mapped to a counterpart of EDUs from the RST-DT. The relatively small number
of matches can be attributed to the fact that the text spans we compare have very
different features. While the goal of our TS approach is to generate well-formed
syntactically simplified sentences, the EDUs in the RST-DT are copied verbatim
from the source at clausal or phrasal level, resulting in an output of varied length
that is usually not grammatically sound. Moreover, in many cases, the EDUs mix
multiple semantic units, whereas our approach aims to split the input into atomic
components, with each of them expressing a coherent and indivisible proposition.
Some characteristic examples are given in Table 15.13. As we are primarily interested in determining whether the constituency and relation labels that are assigned
by our approach are correct, we will focus on the metric of precision in the following,
measuring its ability to find the right markers.[1]

---

[1] The fraction of labels that are successfully retrieved (i.e. recall) is of minor importance in
our setting. In addition, this score might be biased, since a large proportion of EDUs from the
RST-DT corpus is not mapped to a counterpart of simplified propositions in our experiments.
Therefore, we refrain from reporting recall scores.

Table 15.12 Comparison of the properties of the simplified sentences generated by our discourse-aware TS approach and the EDUs of the RST-DT

	# spans	average span length
DISSIM	24,056 simplified sentences	10.5
RST-DT	21,789 EDUs	8.1

Table 15.12 provides an overview of the features of the simplified sentences generated by our discourse-aware TS approach and compares them to the properties of the EDUs from the RST-DT. The number of split sentences produced by our TS framework on the source sentences from the RST-DT is only marginally higher than the number of EDUs contained in it. Furthermore, with an average length of 10.5 tokens, the simplified sentences are slightly longer than the EDUs.

On the basis of the 1,827 matched sentence pairs, we evaluated the performance of our discourse-aware TS approach with regard to the constituency type classification task and the rhetorical relation identification step. The outcome of these examinations will be presented below.

15.2.1.1 Constituency Type Classification

In 88.88% of the matched sentence pairs, the hierarchical relationship that was allocated between a pair of simplified sentences by our reference TS implementation DISSIM corresponds to the nuclearity status of the aligned EDUs from RST-DT, i.e. in case of a nucleus-nucleus relationship in RST-DT, both output sentences from DISSIM are assigned to the same context layer, while in case of a nucleus-satellite relationship the sentence mapped to the nucleus EDU is allocated to the context layer cl, whereas the sentence mapped to the satellite span is assigned to the subordinate context layer $cl+1$. For an example, see Table 15.14.

The majority of the cases where our TS approach assigns a hierarchical relationship that differs from the nuclearity in the RST-DT corpus can be attributed to relative clauses. In RST-DT, given a sentence that contains a relative clause, the EDUs are typically assigned a nucleus-nucleus relationship. In contrast, we regard the information contained in relative clauses as background information that further describes the entity to which it refers. Therefore, we classify a simplified sentence that originates from such a type of clause as a contextual sentence that contributes additional information about its referent contained in the superordinate clause. Table 15.15 provides illustrative examples of this phenomenon.

Table 15.13 Some characteristic examples of simplified sentences generated by our TS framework DISSIM and the corresponding EDUs from the RST-DT

Input	Other major issues hitting highs included American Telephone & Telegraph Co., Westinghouse Electric Corp., Exxon Corp. and Cigna Corp., the big insurer.
DISSIM	• Other major issues hitting highs included American Telephone & Telegraph Co. • Other major issues hitting highs included Westinghouse Electric Corp. • Other major issues hitting highs included Exxon Corp. • Other major issues hitting highs included Cigna Corp. • Cigna Corp. was the big insurer.
RST-DT	• Other major issues • hitting highs • included American Telephone & Telegraph Co., Westinghouse Electric Corp., Exxon Corp. and Cigna Corp., the big insurer.
Input	For both Thomson and British Aerospace, earnings in their home markets have come under pressure from increasingly tight-fisted defense ministries; and Middle East sales, a traditional mainstay for both companies' exports, have been hurt by five years of weak oil prices.
DISSIM	• Earnings in their home markets have come under pressure. • This was from increasingly tight-fisted defense ministries. • This was for both Thomson and British Aerospace. • Middle East sales have been hurt. • This was by five years of weak oil prices. • Middle East sales were a traditional mainstay for both companies' exports.
RST-DT	• For both Thomson and British Aerospace, earnings in their home markets have come under pressure from increasingly tight-fisted defense ministries; • and Middle East sales, a traditional mainstay for both companies' exports, have been hurt by five years of weak oil prices.

Table 15.16 displays the precision that the discourse parser baselines achieve on the 991 sentences of the RST-DT test set in distinguishing between nucleus and satellite spans (*"nuclearity"*). For the approaches in the upper part of the table, the authors report the systems' performance when using gold EDU segmentation, while for those in the lower part the performance is indicated based on automatic segmentation, i.e. when they are fed the output of their respective discourse segmenter.

Table 15.14 Constituency type classification examples

Source	While lawyers arranged individual tie-ups before, the formal network of court reporters should make things easier and cheaper.	
	MAPPED SPAN 1	MAPPED SPAN 2
DISSIM	Lawyers arranged individual tie-ups.	The formal network of court reporters should make things easier and cheaper.
Nuclearity	**context layer 0**	**context layer 1**
RST-DT	While lawyers arranged individual tie-ups before,	the formal network of court reporters should make things easier and cheaper.
Nuclearity	**nucleus**	**satellite**

Table 15.15 Examples of constituency type classifications of sentences that contain a relative clause

Source	The consolidated firm, which would rank among the 10 largest in Texas, would operate under the name Jackson & Walker.	
	MAPPED SPAN 1	MAPPED SPAN 2
DISSIM	The consolidated firm would operate under Jackson & Walker.	The consolidated firm would rank among the 10 largest in Texas.
Nuclearity	**context layer 0**	**context layer 1**
RST-DT	would operate under the name Jackson & Walker.	The consolidated firm, which would rank among the 10 largest in Texas,
Nuclearity	**nucleus**	**nucleus**
Source	Ciba Corning, which had been a 50-50 venture between Basel-based Ciba-Geigy and Corning, has annual sales of about $300 million, the announcement said.	
	MAPPED SPAN 1	MAPPED SPAN 2
DISSIM	Ciba Corning has annual sales of about $ 300 million.	Ciba Corning had been a 50-50 venture between Basel-based Ciba-Geigy and Corning.
Nuclearity	**context layer 0**	**context layer 1**
RST-DT	has annual sales of about $300 million,	Ciba Corning, which had been a 50-50 venture between Basel-based Ciba-Geigy and Corning,
Nuclearity	**nucleus**	**nucleus**

Table 15.16 Precision of DISSIM and the discourse parser baselines on the constituency type classification (*"nuclearity labeling"*) and rhetorical relation identification (*"relation labeling"*) tasks, as reported by their authors. (*) In case of automatic discourse segmentation, for Lin et al. (2019) the F_1 score is available only

	nuclearity	relation
DPLP Ji and Eisenstein (2014)	71.1	61.8
Feng and Hirst (2014)	71.0	58.2
Two-Stage Parser Wang et al. (2017)	72.4	59.7
Lin et al. (2019)	**91.3**	**81.7**
SPADE Soricut and Marcu (2003)	56.1	44.9
HILDA Hernault et al. (2010)	59.7	48.2
PAR- s Joty et al. (2015)	75.2	66.1
Lin et al. (2019)	(86.4)*	(77.5)*
DISSIM (full RST-DT dataset)	**88.9**	**69.5**
DISSIM (RST-DT test set only)	**92.2**	**70.0**

Since our TS framework makes use of the simplified sentences that were generated in the previous step when setting up the semantic hierarchy, it is better comparable to the latter group. The figures show that in this case our approach outperforms all other systems in the constituency classification task by a large margin of 13.7% at a minimum.[2][3]

15.2.1.2 Rhetorical Relation Identification

Table 15.17 displays the frequency distribution of the 19 classes of rhetorical relations that were specified in Taboada and Das (2013) over the RST-DT corpus. The ten most frequently occurring classes make up for 89.45% of the relations that are present in the dataset. We decided to limit ourselves to these classes in the evaluation of the rhetorical relation identification step, with two exceptions.

---

[2] A very recent approach to intra-sentential sentence parsing was proposed in Lin et al. (2019), achieving an F_1 score of 86.4%. However, the authors do not report its precision.

[3] When restricting our experiments to the 38 documents of the test set of the RST-DT corpus, the results (calculated on the basis of 179 mapped sentence pairs) even improve somewhat. The average precision of our reference implementation DISSIM on the nuclearity labeling task slightly increases to 92.2%, surpassing the baseline approaches by at least 17.0%. However, we decided to focus in our analysis on the full dataset, including both the training and test set of the RST-DT, as we consider it to be more meaningful due to its larger size.

Table 15.17 Frequency distribution of the 19 classes of rhetorical relations that are distinguished in Taboada and Das (2013) over the RST-DT corpus *(left)*, and the precision of DisSim's rhetorical relation identification step *(right)*. Note that the *Spatial* relation is not included in the RST-DT gold annotations. Therefore, it is not possible to include an analysis of the performance of our approach with respect to this type of rhetorical relation

RHETORICAL RELATION	COUNT	PERCENTAGE	PRECISION
Elaboration	7675	25.65%	0.5550
Joint	7116	23.78%	0.6673
Attribution	2984	9.97%	0.9601
Same-unit	2788	9.32%	–
Contrast	1522	5.09%	0.7421
Topic-change	1315	4.39%	–
Explanation	966	3.21%	0.7037
Cause	754	2.52%	
Temporal	964	3.22%	0.7895
Background	897	2.30%	0.4459
Evaluation	589	1.97%	
Enablement	546	1.82%	(0.5766)
Comparison	433	1.45%	
Textual organization	364	1.22%	
Condition	317	1.06%	(0.7429)
Topic-comment	255	0.85%	
Summary	220	0.74%	
Manner-means	218	0.73%	
Span	1	0.00%	
	$\sum$ 29,924	$\sum$ 100%	avg.: 0.6948

First, we did not take into account two of the classes included in this set, namely "Topic-change" and "Same-unit". The former encompasses relations that connect large sections of text when there is an abrupt change between topics Schrimpf (2018). Accordingly, this type of relation is relevant when examining larger paragraphs, but of no importance when considering intra-sentential relationships only, as is the case in our approach. The latter is not a true coherence relation. In the RST-DT, it is used to link parts of units separated by embedded units or spans Taboada and Das

Table 15.18 Result of the rhetorical relation identification step on example sentences

Source	Mr. Volk, 55 years old, succeeds Duncan Dwight, who retired in September.	
	MAPPED SPAN 1	MAPPED SPAN 2
DISSIM	Volk succeeds Duncan Dwight.	Duncan Dwight retired in September.
RST-DT	Mr. Volk, 55 years old, succeeds Duncan Dwight,	who retired in September.
Rhet. rel.	**Elaboration**	
Source	Three seats currently are vacant and three others are likely to be filled within a few years, so patent lawyers and research-based industries are making a new push for specialists to be added to the court.	
	MAPPED SPAN 1	MAPPED SPAN 2
DISSIM	Three seats are vacant.	Three others are likely to be filled within a few years.
RST-DT	Three seats currently are vacant	and three others are likely to be filled within a few years,
Rhet. rel.	**Joint**	
Source	Mr. Carpenter notes that these types of investors also are "sophisticated" enough not to complain about Kidder's aggressive use of program trading.	
	MAPPED SPAN 1	MAPPED SPAN 2
DISSIM	These types of investors are "sophisticated" enough not to complain about Kidder's aggressive use of program trading.	This is what Carpenter notes.
RST-DT	that these types of investors also are "sophisticated" enough not to complain about Kidder's aggressive use of program trading.	Mr. Carpenter notes
Rhetorical relation	**Attribution**	

(2013).[4] Second, similar to Benamara and Taboada (2015), we merged the two highly related classes of "Cause" and "Explanation" into a single category, since

---

[4] For example, in the sentence *"The petite, 29-year-old Ms. Johnson, dressed in jeans and a sweatshirt, is a claims adjuster with Aetna Life Camp Casualty.", a "Same-unit" relation holds between the two underlined text spans* Kibrik and Krasavina (2005).

they both indicate a causal relationship. Consequently, we ended up with a set of seven rhetorical relations, aggregating 75.74% of the rhetorical relations included in the RST-DT.

Some examples of rhetorical relations that are assigned between pairs of decomposed sentences by our TS framework DISSIM are illustrated in Table 15.18. The right column in Table 15.17 displays the precision of our TS approach for each class of rhetorical relation when run over the sentences from the RST-DT.[5] With a score of 96.01%, the "Attribution" relation reaches by far the highest precision. The remaining relations, too, show decent scores, with a precision of around 70%. The only exception is the "Background" relation, showing a precision of 44.59%. The difficulty with this type of relationship is that it signifies a very broad category that is not signalled by discourse markers and therefore hard to detect by our approach Taboada and Das (2013). Table 15.19 illustrates some example sentences.

With an average precision of 69.5% in the relation labeling task (see Table 15.16, "relation"),[6] our approach again surpasses all the discourse parser baselines under consideration when using automatic discourse segmentation.[7] [8]

Figure 15.1 illustrates the distribution of the rhetorical relations that were identified by our discourse-aware TS approach on the source sentences from the RST-DT. When comparing it to that of the manually annotated gold relations displayed in Table 15.17, it turns out that there is a very high similarity between the two of them. In both cases, "Joint" and "Elaboration" are by far the most frequently occurring rhetorical relations, though in reversed order, followed by the relations of "Contrast" and "Attribution", which again swap places in the ranking. Note that the "Temporal"

---

[5] These scores again refer to the results obtained on the full RST-DT corpus, including both training and test set. When limiting our analysis to the 38 documents of the test set, the results only change slightly, leading to an average precision of 70.04%. However, we decided to focus on the full RST-DT dataset in the relation labeling task, too, since we consider it to be more meaningful for our purposes due to its larger size.

[6] The average precision refers to the scores of the selected subset of rhetorical relations only, i.e. the ones that are printed in bold in Table 15.17.

[7] with the exception of Lin et al. (2019)'s parser, for which only the F_1 score is reported by the authors, though. Hence, it is not directly comparable to the other approaches whose performance is analyzed based on their precision.

[8] Note that the baselines' precision scores are computed based on the full set of 19 relation classes that are used for the annotation of the RST-DT. Consequently, their average precision scores are not directly comparable to the score achieved by our approach, as they do not include exactly the same set of rhetorical relations. However, as the relations that are not considered in the evaluation of our approach make up for less than a quarter of the relations occurring in the RST-DT corpus, we assume that it still allows for a reliable approximation of the performance of the baselines as compared to the approach we propose.

Table 15.19 Examples of sentences where a "Background" relation is assigned by DISSIM, differing from RST-DT's gold rhetorical relation. However, the background relationship is valid in those cases as well

Source	In addition, economists are forecasting a slowdown in foreign direct investments as businessmen become increasingly wary of China's deteriorating political and economic environment.		
	SPAN 1	SPAN 2	RHETORICAL RELATION
DISSIM	Economists are forecasting a slowdown in foreign direct investments.	Businessmen become increasingly wary of China's deteriorating political and economic environment.	**Background**
RST-DT	In addition, economists are forecasting a slowdown in foreign direct investments	as businessmen become increasingly wary of China's deteriorating political and economic environment.	**Temporal**
Source	Meanwhile, analysts said Pfizer's recent string of lackluster quarterly performances continued, as earnings in the quarter were expected to decline by about 5%.		
	SPAN 1	SPAN 2	RHETORICAL RELATION
DISSIM	Pfizer's recent string of lackluster quarterly performances continued.	Earnings in the quarter were expected to decline by about 5%.	**Background**
RST-DT	Pfizer's recent string of lackluster quarterly performances continued,	as earnings in the quarter were expected to decline by about 5%.	**Elaboration**

relation is much more common in the output produced by our TS approach than in the ground truth. Also the "Spatial" relation, which we introduced to capture local relationships and which is not part of the current set of rhetorical relations included in RST, appears quite often in the semantic hierarchy generated by our framework. The remaining classes, "Enablement", "Background", "Cause" and "Condition", are less frequent, though at a rate comparable to the gold relations' distribution.

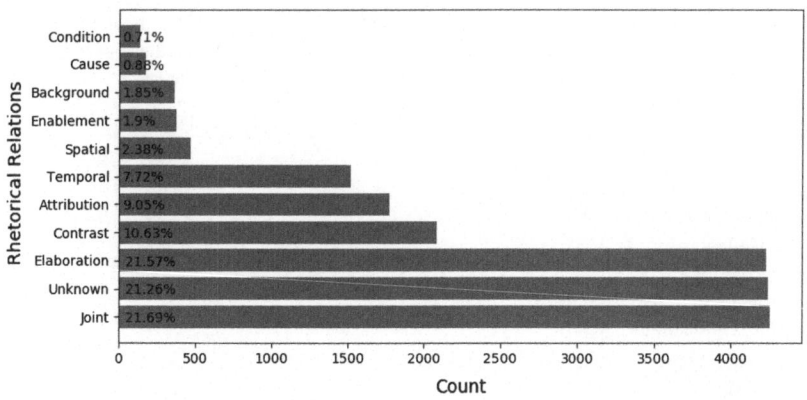

Figure 15.1 Distribution of the rhetorical relations identified by our discourse-aware TS approach on the RST-DT

It must be noted that in about a fifth of the cases, our TS approach is not able to identify a rhetorical relation between a pair of decomposed sentences ("Unknown"). For the most part, this can be attributed to sentence pairs whose relation is not explicitly stated in the underlying source sentence. Since our approach is based on cue phrases, searching for discourse markers that explicitly signal rhetorical relations, it has difficulties in identifying relations that can merely be implied. In the future, we aim to address this gap, using a more sophisticated approach that is able to also capture implicit relationships between simplified sentences, e.g. by following the work of Feng and Hirst (2014), Joty et al. (2015), Liu and Lapata (2017) and Lin et al. (2019) that apply supervised ML methods that are based on a rich set of features to complement the lookup of cue phrases.

15.2.2 Manual Analysis

Below, we report our findings of the manual analysis addressing the performance of our discourse-aware TS approach with regard to the semantic hierarchy established between the split sentences.

Table 15.20 Results of the manual analysis regarding the quality of the semantic hierarchy that is established by our proposed discourse-aware TS approach between the split sentences. We report the inter-annotator agreement score for each category using Fleiss' κ

Category	Yes	No	Malformed	Unspecified	Fleiss' κ
Limitation to core information	**0.6815**	0.2000	0.1185	–	0.39
Soundness of the contextual proposition	**0.8312**	0.1688	–	–	0.51
Correctness of the context allocation	**0.9318**	0.0682	–	–	0.41
Properness of the semantic relationship	**0.6984**	0.0700	–	0.2316	0.69

The results of the human evaluation are displayed in Table 15.20. The inter-annotator agreement was calculated using Fleiss' κ Fleiss et al. (1971).[9] The obtained rates were 0.39, 0.51, 0.41 and 0.69 for the four categories that we assessed. Hence, the figures indicate fair to substantial agreement between the three annotators, suggesting that the scores we got for the four categories under consideration present a reliable result. According to the agreement rates, the limitation to core information shows the highest level of subjectivity and was the most difficult to judge, leading to a greater divergence between the annotators as compared to the other three dimensions. Conversely, determining the properness of the semantic relationship is the easiest task, achieving by far the highest level of agreement. However, this is partly due to the comparatively high percentage of sentence pairs for which our TS approach did not specify a rhetorical relation (21.89%), allowing for an unambiguous answer.[10]

The final scores for the four manually analyzed categories were calculated by averaging over the annotators' scores for the simplified propositions of the 100 randomly sampled sentences, providing very promising results.

---

[9] Unlike before in Section 15.1.2.1, where we used Cohen's κ to determine the agreement between the annotators, we now apply Fleiss κ, since the latter works for any number of raters, while the former score is suitable only in case of exactly two raters.

[10] Without considering the "unspecified" class, the agreement rate for this category considerably drops to 0.3246, indicating a fair agreement between the annotators.

Limitation to core information   In more than two out of three cases, the annotators marked the propositions that were classified as core sentences by our discourse-aware TS approach as correct, thus approving that they have a meaningful interpretation and that their content is truly restricted to core information of the underlying source sentence. Only about 12% of the simplified sentences are malformed according to our annotations. The remaining fifth of output core sentences was judged as being misclassified, i.e. they rather contribute less relevant background data than key information of the input.

Soundness of the context proposition   When considering the soundness of the output propositions that were classified as contextual sentences, we reach a very high score. Only about 17% of the sentences were labelled as malformed, while as many as 83% present accurate contextual propositions, expressing a meaningful context fact that is asserted by the input and can be properly interpreted.

Correctness of the context allocation   Regarding the question of the correctness of the context allocation, we achieve similarly good results in our analysis. 93% of the context sentences are assigned to their respective parent sentence, whereas only 7% of them are misallocated, according to the annotators' labels. Note that only those propositions that were flagged as being sound in the previous step by the corresponding annotator were considered here, reducing the set of simplified context sentences taken into account by 15.7%, 15.4% and 19.6%, respectively.[11]

Properness of the semantic relationship   Finally, our evaluation revealed that our proposed TS approach shows a decent performance for the rhetorical relation identification step, too. More than two-thirds of the sentence pairs are classified with the correct rhetorical relation, according to our manual analysis. Only 7% of them are assigned an improper relation. However, in nearly a quarter of the cases, our TS approach was not able to identify a semantic relationship between a given pair of sentences. This can be explained by the fact that for this subtask, our framework follows a rather simplistic approach that is based on cue phrases. Therefore, it fails to identify a semantic relationship whenever none of the specified keywords[12] appears in the underlying input sentence. As a result, this approach provides very precise results. Covering only a small subset of rhetorical relations (see Section 15.2.1.2),

---

[11] For the computation of the inter-annotator agreement rate of this category, we only incorporate the 218 contextual propositions that all three annotators previously labelled as a sound output sentence, expressing an informative and meaningful context fact.

[12] The interested reader may refer to Section 20.1 in the online supplemental material for the full set of cue phrases and corresponding rhetorical relations.

it lacks in completeness, though. Note that here, too, we make use of the results of the previous step, considering only those propositions that were labelled as being allocated to the correct parent sentence. In that way, the set of context sentences taken into account slightly decreases by 14.1%, 11.8% and 3.3%, respectively.[13]

To sum up, the scores from our manual analysis indicate that our proposed discourse-aware TS approach shows a very good performance in establishing a contextual hierarchy between the split sentences by first distinguishing core sentences that contain the main information of the input from contextual sentences, whose content discloses less relevant background information, and then allocating each contextual proposition to its corresponding parent sentence. The results obtained in the second subtask, the rhetorical relation identification step, are satisfactory, too. However, we observed some room for improvement here, in particular with respect to the semantic relationships that are not explicitly expressed in the input text in the form of cue phrases.

15.3   Subtask 3: Extracting Relations and their Arguments

In addition to the intrinsic evaluation of our proposed discourse-aware TS approach that was described above, we carried out an extrinsic evaluation. Its goal is to predict the merits of the representation of complex sentences in the form of a semantic hierarchy of syntactically simplified minimal propositions for downstream state-of-the-art Open IE systems.

15.3.1 Performance of the Reference Open Information Extraction Implementation Graphene

In the following, we report the results of the automatic comparative evaluation of the performance of our Open IE reference implementation Graphene and the Open IE baseline systems, as well as the findings of the manual analyses performed on their output.

---

[13] As before, for the agreement score we only consider the 247 contextual propositions that all three annotators previously flagged as being correctly assigned to their respective parent sentence.

15.3.1.1 Automatic Comparative Analysis of the Outputs

In order to compare our Open IE reference implementation Graphene to existing Open IE approaches, we conducted an automatic evaluation on three recently proposed Open IE benchmarks: OIE2016, CaRB and WiRe57 (see Sections 2.3.2 and 14.3.2). Each of them comes with an evaluation framework that allows for a comparison of the most established Open IE systems on the basis of their precision and recall scores.

OIE2016   Figure 15.2 shows the PR curves of Graphene and the Open IE baseline systems listed in Section 14.3.1 on the OIE2016 benchmark. Their precision, recall, F_1 and AUC scores are given in Table 15.21.

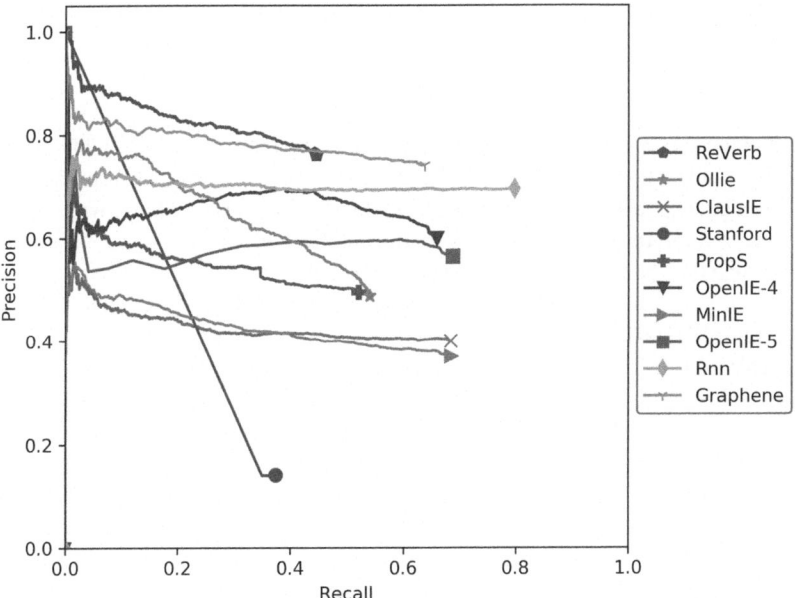

Figure 15.2 Comparative analysis of the performance of our reference implementation Graphene against baseline Open IE systems on the OIE2016 benchmark

The results reveal that with an average precision of 78.5%, Graphene succeeds in extracting highly accurate relational tuples at a high recall rate of 63.8%. Of all the Open IE systems under consideration, REVERB (83.6%) is the only approach

Table 15.21 Average precision, recall, F^1 and AUC scores of a variety of state-of-the-art Open IE approaches on the OIE2016 benchmark

System	Precision	Recall	F_1	AUC
REVERB	**0.836**	0.446	0.582	0.296
OLLIE	0.660	0.541	0.595	0.357
ClausIE	0.430	0.684	0.528	0.294
Stanford Open IE	0.140	0.375	0.204	0.203
PropS	0.556	0.521	0.538	0.290
OpenIE-4	0.661	0.660	0.660	0.434
MinIE	0.431	0.686	0.529	0.296
OpenIE-5	0.581	0.688	0.630	0.399
RnnOIE	0.701	**0.799**	**0.747**	**0.559**
Graphene	0.785	0.638	0.704	0.502

that outperforms Graphene in terms of precision, though at the cost of a low recall score of only 44.6%. RnnOIE (70.1%), OpenIE-4 (66.1%) and OLLIE (66.0%) also show a good precision, following on places three to five in the ranking. With regard to recall, RnnOIE (79.9%) is the best-performing system, setting the bar very high. Although Graphene does not achieve this target value, it is able to compete with the other high-precision Open IE systems, including OpenIE-4 (66.0%), OLLIE (54.1%) and OpenIE-5 (68.8%).

This lack in recall is no surprise, since Graphene is not designed to extract fine-grained relations for each and every verb contained in an input sentence Cetto (2017), as opposed to the nature of the OIE2016 gold tuples (see Section 2.3.2). Rather, Graphene determines the main relations included in the source sentence, together with attached contextual arguments that often hold some of these verbal expressions. In an in-depth analysis of the output produced by Graphene, we found that only 33.72% of the unmatched gold extractions were caused by a wrong argument assignment. In the majority of the cases (66.28%), Graphene did not extract tuples with a matching relational phrase. In a further investigation, we figured out that in 56.80% of such cases, the relational phrase of a missed gold standard extraction was actually contained in one of the argument phrases, e.g.:

"Japan may be a tough market for outsiders to penetrate, and the U.S. is hopelessly behind Japan in certain technologies."

- Unmatched gold extraction: ⟨*outsiders*; *may penetrate*; Japan⟩
- Tuples extracted by Graphene:

1. ⟨*Japan; may be; a tough market for outsiders to penetrate* ⟩
2. ⟨ *the U.S.; is hopelessly behind Japan; in certain technologies*⟩

This number even grows to 67.89% when considering the lemmatized form of the head, e.g.:
   "The seeds of 'Raphanus sativus' can be pressed to extract radish seed oil."

- Unmatched gold extraction: ⟨*radish seed oil; can be extracted; from the seeds*⟩
- Tuples extracted by Graphene:

1. ⟨*the seeds of 'Raphanus sativus' can be pressed; to extract radish seed oil*⟩

Another 5.22% of unmatched relational phrases would have been recognized by Graphene if they were compared based on their lemmatized head.

Regarding F_1, with a score of 70.4%, Graphene closely follows the best-performing system RnnOIE (74.7%), with OpenIE-4 (66.0%) and OpenIE-5 (63.0%) in third and fourth place. Finally, when considering the AUC score as a combined measure of precision and recall, Graphene (50.2%) again is the second best-performing system after RnnOIE (55.9%), once again followed by OpenIE-4 (43.4%) and OpenIE-5 (39.9%).

Table 15.22 reports a number of descriptive statistics on the relations extracted from the 3,200 sentences of the OIE2016 benchmark, providing an insight into the main characteristics of the output produced by the various Open IE baseline approaches. The numbers show that with more than 20k extractions in total and six extractions per input sentence on average, the Stanford Open IE system extracts by far the highest number of relations from the source text, followed by MinIE (4.43) and ClausIE (4.36). However, a fifth of its extractions are redundant, i.e. they contain subsequences of other tuples that were already extracted. REVERB, on the other hand, makes very conservative choices, leading not only to a high precision (see Table 15.21) and no redundancy in the extractions, but also to a low coverage rate of only 42%, indicating that not even half of the input tokens are contained in the extracted relations. These findings are indeed supported by the high number of source sentences from which REVERB does not produce any extraction (559 in total, accounting for 17% of the input sentences), as well as the low average triple length of 7.53 tokens. These scores are only undercut by Stanford Open IE (41% and 7.20). At the other end of the scale, PropS can be found. Its extractions present

Table 15.22 Descriptive statistics on the relations extracted from the 3,200 sentences of the OIE2016 corpus. We report scores for a number of basic statistics, including the total number of relational tuples extracted from the input (#E); the average number of facts extracted per source sentence (#E/S); the percentage of non-redundant extractions (%E$_{non-redundant}$); the number of sentences from which not even one fact is extracted (# S<1E); the average length of the extracted relational tuples on the basis of the average number of tokens included in the phrases in subject, predicate and object position; and the percentage of tokens from the input that are contained in at least one of the extracted relational tuples (coverage). (*) calculated over the 1,769 sentences from OIE2016's test set

System	#E	#E/S	%E$_{non-redundant}$	#S<1E	avg. triple length (s, p, o)	coverage
Gold standard*	4,852	2.74	99.94%	0	13.43 (3.39, 1.39, 8.65)	0.74
REVERB	4,176	1.31	**100.00%**	559	7.53 (2.27, 2.50, 2.77)	0.42
OLLIE	8,226	2.57	98.72%	318	9.66 (2.55, 3.57, 3.54)	0.59
ClausIE	13,946	4.36	88.06%	18	11.49 (2.70, 1.45, 7.34)	0.85
Stanford Open IE	**20,527**	**6.41**	80.98%	507	**7.20** (1.61, 1.90, 3.70)	0.41
PropS	7,871	2.46	99.87%	48	16.67 (1.99, 1.04, 13.65)	**0.86**
OpenIE-4	9,236	2.89	99.89%	91	11.25 (2.98, 2.98, 5.29)	0.77
MinIE	14,175	4.43	98.15%	43	8.11 (2.14, 3.26, 2.71)	0.62
OpenIE-5	9,472	2.96	98.21%	87	12.43 (2.82, 1.74, 7.87)	0.62
RnnOIE	8,575	2.68	99.84%	7	12.12 (2.67, 1.89, 7.56)	0.69
Graphene	6,242	1.95	99.82%	165	11.97 (2.78, 2.10, 7.08)	0.73
Average	10,244.60	3.20	96.35%	184.30	10.84 (2.45, 2.24, 6.15)	0.66

the highest average triple length (16.67); notably with 13.65 tokens on average, the arguments in object position are comparatively long. This can be attributed to the fact that PropS is one of the few Open IE systems under consideration that do not only produce binary relations, but instead aims to generate n-ary relations with more than one argument in the object slot (see Table 15.23). Since for the purpose of this analysis, we simply join them into a contiguous argument phrase, we typically end up with relatively long object phrases. Moreover, PropS succeeds in producing very complete results, covering 86% of the input tokens in the extracted relations.

With 6,242 extractions, our reference Open IE implementation Graphene outputs the second lowest number of relational tuples, right after REVERB. However, unlike the latter, it generates a large percentage of n-ary relations (see Table 15.23), resulting in a decent coverage rate of 73%. With 165 sentences, it fails to produce an extraction

Table 15.23 Distribution of the arity of the extracted relations on the 3,200 sentences of the OIE2016 dataset. (*) calculated over the 1,769 sentences from OIE2016's test set

System	unary	binary	ternary	quaternary	≥ quaternary
Gold standard*	8%	52%	30%	8%	1%
REVERB	–	100%	–	–	–
OLLIE	–	100%	–	–	–
ClausIE	–	100%	–	–	–
Stanford Open IE	–	100%	–	–	–
PropS	1%	35%	34%	16%	14%
OpenIE-4	5%	95%	–	–	–
MinIE	3%	97%	–	–	–
OpenIE-5	6%	65%	24%	5%	1%
RnnOIE	11%	66%	21%	2%	–
Graphene	12%	56%	24%	6%	1%

for about 5% of the input sentences. Furthermore, practically none of its extractions are redundant (0.18%) and the average triple length of its relations (11.97) is close to the average length of all the tested Open IE systems' extractions.

Table 15.23 provides an overview of the arity of the relations extracted by the different Open IE approaches from the 3,200 sentences of the OIE2016 dataset. Though only about half of the ground truth tuples are binary, the majority of the Open IE systems are limited to extracting relations with exactly two argument slots, including REVERB, OLLIE, ClausIE,[14] Stanford Open IE, OpenIE-4 and MinIE. On the contrary, PropS, OpenIE-5, RnnOIE and Graphene support the extraction of n-ary relations, as provided in the gold standard dataset. While for OpenIE-5, RnnOIE and Graphene, the tuples with an arity higher than two make up for about a quarter to a third of the extractions, the output generated by PropS even includes almost two third of non-binary relations.

CaRB Figure 15.3 illustrates the PR curves of Graphene and the Open IE baseline approaches on the CaRB benchmark when using the ground truth reference tuples from the OIE2016 dataset in combination with the improved evaluation metrics proposed in the CaRB framework (see Section 14.3.3.1). The corresponding precision,

---

[14] In our analyses, we used the default configuration of ClausIE, which is limited to extracting binary relations. However, it can also be customized to extract n-ary relations.

recall and F_1 scores (at the maximum F_1 point), as well as the AUC scores are listed in Table 15.24.

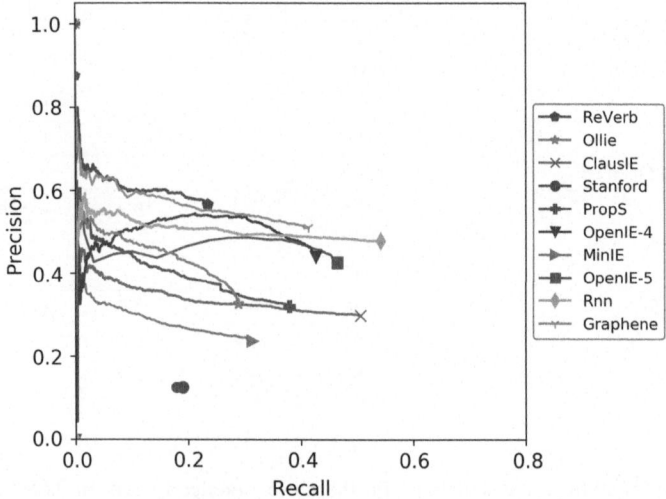

Figure 15.3 Comparative analysis of the performance of our reference implementation Graphene against baseline Open IE systems on the CaRB benchmark when using the gold tuples from the OIE2016 dataset in combination with the improved evaluation metrics from the CaRB framework

The results show that the overall precision and recall scores decrease due to the more restrictive matching function that, amongst other things, penalizes extractions for misidentifying parts of a relation in an argument slot (or vice versa) and leads to reduced precision scores when there are several very similar and redundant extractions or when the argument phrases of the extracted relations are overly long. However, there is no significant change in the order of the tested Open IE systems as compared to the OIE2016 benchmark framework. REVERB (56.7%) is still the best-performing system with respect to precision, while RnnOIE surpasses all the systems under consideration with regard to recall (54.3%), F_1 (50.8%) and AUC (27.4%). Graphene follows in second place when considering precision (51.1%), F_1 (45.8%) and AUC (23.6%). Regarding recall (41.4%), Graphene again does not reach the high bar set by RnnOIE, yet it is able to compete with the other high-precision Open IE systems, such as OpenIE-5 (45.4%) and OpenIE-4 (41.6%), as was the case before on the OIE2016 benchmark.

Table 15.24 Precision, recall, F_1 (at maximum F_1 point) and AUC of a variety of state-of-the-art Open IE approaches on the CaRB benchmark when using the gold tuples from the OIE2016 dataset in combination with the improved evaluation metrics from the CaRB framework

System	Precision	Recall	F_1	AUC
REVERB	**0.567**	0.235	0.332	0.144
OLLIE	0.363	0.276	0.314	0.130
ClausIE	0.301	0.502	0.376	0.172
Stanford Open IE	0.125	0.190	0.151	0.102
PropS	0.324	0.378	0.349	0.153
OpenIE-4	0.464	0.416	0.439	0.214
MinIE	0.245	0.303	0.271	0.090
OpenIE-5	0.445	0.454	0.449	0.218
RnnOIE	0.478	**0.543**	**0.508**	**0.274**
Graphene	0.511	0.414	0.458	0.236

WiRe57   As the previous analyses conducted on the OIE2016 and CaRB benchmarks were based on the same set of source sentences and ground truth tuples, differing only in their evaluation methods, we additionally consider a third dataset, the WiRe57 Open IE benchmark, to examine whether the previously achieved results can be transferred to other corpora.

Table 15.25 reports the token-level precision, recall and F_1 scores of our reference implementation Graphene and the Open IE baselines listed in Section 14.3.1. As before, REVERB (56.9%) outperforms all tested systems with regard to precision, followed by OpenIE-4 (50.1%) and Graphene (41.9%) in second and third place. While on the OIE2016 and CaRB benchmarks, RnnOIE surpassed all the other Open IE approaches under consideration in terms of recall and the overall F_1 score, it is now the system that performs worst with respect to all three evaluation metrics (19.0%, 11.0%, 13.9%). This may be an indication that this approach is specially geared to the needs of the OIE2016 gold standard tuples and, consequently, the results obtained for RnnOIE on the previous two benchmarks cannot be transferred to other datasets.

Conversely, MinIE, which was among the systems with the lowest overall performance on the two previously used Open IE benchmarks, beats all tested approaches on the WiRe57 dataset when considering recall (32.3%) and F_1 (35.8%), with

Table 15.25 Token-level precision, recall and F_1 of a variety of state-of-the-art Open IE approaches on the WiRe57 benchmark

System	Precision	Recall	F_1
REVERB	**56.9%**	12.1%	0.200
OLLIE	37.4%	17.5%	0.239
ClausIE	40.1%	29.8%	0.342
Stanford Open IE	21.0%	18.8%	0.198
PropS	22.2%	16.2%	0.187
OpenIE-4	50.1%	18.2%	0.267
MinIE	40.0%	**32.3%**	**0.358**
OpenIE-5	31.5%	20.7%	0.250
RnnOIE	19.0%	11.0%	0.139
Graphene	41.9%	21.1%	0.281

ClausIE (29.8% and 34.2%) and Graphene (21.1% and 28.1%) following in the ranking.

Table 15.26 illustrates a number of descriptive statistics on the relations extracted from the 57 sentences of the WiRe57 benchmark. The figures show that once again, with 371 extractions in total and 6.51 extractions per sentence on average, Stanford Open IE is the system that extracts the largest number of relations from the input, as it has already been the case before on the OIE2016 benchmark (see Table 15.22). On the contrary, both OpenIE-5 and RnnOIE only generate a single relational tuple per source sentence, which can be attributed to the fact that they are the only approaches (together with Graphene) whose output is not limited to the extraction of binary relations, but presents a high percentage of relations with a higher arity (see Table 15.27). Thus, they tend to merge relations into a single tuple where the other Open IE approaches under consideration instead produce several extractions that are based on the same relational phrase, with each of them combined with a different set of arguments.

Moreover, analogous to our findings on the OIE2016 benchmark, all tested Open IE systems hardly generate any redundant extractions, with the notable exception of Stanford Open IE, which outputs more than 20% of relations whose subsequences are already contained in one of its previously extracted tuples. Furthermore, all approaches perform well in extracting at least one relation per output sentence. PropS is the only system that fails to produce an extraction for a considerable number of input sentences (14 in total, accounting for 25% of the source sentences).

Table 15.26 Descriptive statistics on the relations extracted from the 57 sentences of the WiRe57 corpus. We report scores for a number of basic statistics, including the total number of relational tuples extracted from the input (#E); the average number of facts extracted per source sentence (#E/S); the percentage of non-redundant extractions ($\%E_{non-redundant}$); the number of sentences from which not even one fact is extracted (#S<1E); the average length of the extracted relational tuples on the basis of the average number of tokens included in the phrases in subject, predicate and object position; and the percentage of tokens from the input that are contained in at least one of the extracted relational tuples (coverage)

System	#E	#E/S	$\%E_{non-redundant}$	#S<1E	avg. triple length (s, p, o)	coverage
Gold standard	343	6.02	98.83%	0	9.75 (3.03, 1.80, 4.92)	0.72
REVERB	79	1.39	**100.00%**	7	7.46 (2.27, 2.63, 2.56)	0.47
OLLIE	145	2.54	99.31%	6	10.06 (2.47, 3.88, 3.70)	0.57
ClausIE	223	3.91	**100.00%**	**0**	12.35 (2.59, 1.36, 8.40)	**0.81**
Stanford Open IE	**371**	**6.51**	79.51%	7	7.43 (1.61, 1.71, 4.11)	0.41
PropS	184	3.23	98.91%	14	10.76 (3.36, 1.05, 6.34)	0.62
OpenIE-4	101	1.77	**100.00%**	2	12.63 (2.90, 1.48, 8.26)	0.74
MinIE	252	4.42	99.21%	1	**7.42** (2.08, 3.04, 2.31)	0.61
OpenIE-5	57	1.00	**100.00%**	**0**	13.19 (2.65, 1.46, 9.09)	0.60
RnnOIE	57	1.00	**100.00%**	**0**	11.75 (2.26, 1.70, 7.79)	0.52
Graphene	133	2.33	**100.00%**	**0**	12.44 (2.63, 1.87, 7.93)	0.78
Average	160.20	2.81	97.69%	3.70	10.55 (2.48, 2.02, 6.05)	0.61

The relations output by MinIE (7.42), Stanford Open IE (7.43) and REVERB (7.46) present the shortest average triple length, while OpenIE-5 (13.19) produces the longest extractions, resulting again from the fact that more than half of its relational tuples are ternary and quaternary. Finally, while the extractions produced by Stanford Open IE and REVERB only cover 41% and 47% of the input tokens, respectively, ClausIE succeeds in generating very complete extractions (achieving a coverage rate of 81%), closely followed by Graphene, whose extractions include 78% of the tokens from the source sentences.

Table 15.27 outlines the distribution of the arity of the relations extracted by the different Open IE systems on the 57 sentences of the WiRe57 benchmark. As it has been the case before on the OIE2016 dataset, the extractions produced by PropS, OpenIE-5, RnnOIE and Graphene are not restricted to binary relations. Instead, a high percentage of tuples that they extract comprise three or four (or even more) arguments. What stands out here is that with 36% to 57%, the proportion of n-ary

Table 15.27 Distribution of the arity of the relations extracted from the 57 sentences of the WiRe57 dataset

System	unary	binary	ternary	quaternary	≥ quaternary
Gold standard	–	74%	26%	–	–
REVERB	–	100%	–	–	–
OLLIE	–	100%	–	–	–
ClausIE	–	100%	–	–	–
Stanford Open IE	–	100%	–	–	–
PropS	7%	43%	22%	13%	14%
OpenIE-4	7%	93%	–	–	–
MinIE	2%	98%	–	–	–
OpenIE-5	2%	42%	46%	11%	–
RnnOIE	–	53%	44%	3%	–
Graphene	5%	60%	26%	7%	3%

relations is even higher as compared to the output produced from the sentences of the OIE2016 benchmark.

Summary The automatic comparative analysis conducted on the OIE2016, CaRB and WiRe57 Open IE benchmarks revealed that our reference Open IE implementation Graphene extracts a large number of correct tuples from complex input sentences. When considering both F_1 and AUC scores, it is consistently among the top three systems (second place for OIE2016 and CaRB, third place for WiRe57), underpinning the high quality of its output with regard to the coverage and accuracy of the relations it extracts.

15.3.1.2 Error Analysis

Examining by hand the 414 relations that were extracted by Graphene from the randomly sampled 200 sentences of the OIE2016 Open IE benchmark, we identified five major types of errors:[15]

---

[15] In total, 58.94% of the extracted relational tuples were annotated as being accurate, i.e. they express a meaningful statement that is asserted by the input and neither over- nor under-specified.

Table 15.28 Major types of errors identified in the relations extracted by our reference Open IE implementation Graphene

ERROR	INPUT	OUTPUT
overly long argument phrase	Spennymoor Town F.C. are the main local football team and won the FA Carlsberg Vase, after winning 2-1 in the final at Wembley Stadium against Tunbridge Wells in May 2013.	⟨*Spennymoor Town F.C.; was winning*; **in the final at Wembley Stadium against Tunbridge Wells in May 2013**⟩
	This change was soon picked up by Huguenot writers, who began to expand on Calvin and promote the idea of the sovereignty of the people, ideas to which Catholic writers and preachers responded fiercely.	⟨*Huguenot writers; began;* **to expand on Calvin and promote the idea of the sovereignty of the people**⟩
uninformative extraction	The 1988 tax act created a federal bill of rights spelling out IRS duties to protect taxpayers' rights in the assessment and collection of taxes.	⟨**Rights;** *were spelling out; IRS duties*⟩
extracted tuple is not asserted by the source	The APL has often been used by plaintiffs asserting that local governments have illegally taken land or imposed taxes.	⟨**The APL;** *imposed; taxes*⟩
overgenerated predicate	New York City is iconic not only for Americans, but also for many Europeans as the city of melting pot where many ethnic groups live, often in specific neighborhoods, such as Chinatown, Little Italy.	⟨*New York City;* **is iconic not only for Americans, but also for many Europeans as the city of melting pot**⟩
	He has on several occasions been a guest at the White House.	⟨*He;* **has on several occasions been**; *a guest at the White House*⟩
no meaningful interpretation	23.8% of all households were made up of individuals and 13.0% had someone living alone who was 65 years of age or older.	⟨*23.8% of all households; were made up*; **of 13.0% had someone living alone**⟩

- An extraction where the *argument phrase is too long* in the sense that it subsumes other extractions. Hence, the extracted relational tuple does not represent an indivisible item, but rather mixes multiple semantic units that should be further split into atomic components. This type of error represents by far the most prevalent mistake, affecting 22.95% of the extractions included in the sample.
- An *uninformative extraction* where critical information from the input is omitted, resulting in a relational tuple that lacks important pieces of information for a proper interpretation of the output. With 9.42%, this error also occurs relatively frequently in the extracted relations.
- A relation to which the wrong argument was assigned, leading to a statement that is comprehensible, but *not asserted* by the source sentence. This type of mistake applies to 7.73% of the extracted relational tuples in total, according to our examinations.
- An extraction with an *overgenerated predicate*, where the relational phrase is overly specific, conveying too much information to be useful in downstream NLP tasks. This error affects 6.76% of the extractions from our sample.
- An extraction where *no meaningful interpretation* of the relational tuple is possible. With 6.28%, this type of error is the least common among the extracted relations.

Other types of errors occur very rarely, affecting only 0.24% of the relational tuples included in our analysis. For better illustration, each error class is depicted in Table 15.28 by means of an example sentence.

15.3.1.3 Qualitative Analysis of the Extracted Relations

Below, we compare and discuss the characteristics of the output produced by the various Open IE systems that we included in our analyses. Figures 15.4 to 15.12 provide representative examples of the relational tuples extracted by the different approaches, given the following input sentence:

> *"He nominated Sonia Sotomayor on May 26, 2009 to replace David Souter; she was confirmed on August 6, 2009, becoming the first Supreme Court Justice of Hispanic descent."*

Graphene As compared to the Open IE baseline approaches, Graphene returns a smaller number of relational tuples. However, it augments them with additional contextual information (e.g., (1a) and (1b) in Figure 15.4), resulting in an average triple length that is slightly above the average of all the tested systems (see Table 15.22

and Table 15.26). In contrast to Graphene, many of the other Open IE systems under consideration tend to either produce overly long argument phrases that mix multiple semantic units (e.g., (1) and (2) in Figure 15.5) or output tuples that omit critical contextual information, hindering a proper interpretation of the extractions (e.g., (2) in Figure 15.9).

```
Graphene:
(1) He;      nominated;    Sonia Sotomayor
(1a)      S:PURPOSE        to replace David Souter
(1b)      S:TEMPORAL       on May 26, 2009
(2) She;     was;         confirmed
(2a)      S:TEMPORAL       on August 6, 2009
(3) She;    was becoming;  the first Supreme Court Justice of Hispanic
descent
```

Figure 15.4 Relations extracted by Graphene

ClausIE ClausIE shows a tendency to extract overly long argument phrases that do not represent an atomic unit, but instead subsume other self-contained facts (e.g., (1) and (2) in Figure 15.5). In many cases, these arguments are generated by combining and rearranging phrasal components from the input sentence in various different orders.

```
ClausIE:
(1) he;      nominated;    Sonia Sotomayor on May 26 2009 to replace David
    Souter
(2) she;    was confirmed;  on August 6 2009 becoming the first Supreme
Court Justice of Hispanic descent
(3) she;    was confirmed;  becoming the first Supreme Court Justice of
Hispanic descent
```

Figure 15.5 Relations extracted by ClausIE

Stanford Open IE Stanford Open IE, too, has a tendency to generate arguments through phrasal permutation. However, instead of combining multiple separate phrasal components into a single argument phrase, it typically sticks with a given noun phrase and adds or deletes optional elements, such as adjectives or adverbs (e.g., (3), (5), (6) and (7) in Figure 15.6). Therefore, Stanford Open IE produces not only the tuples with the shortest average triple length, but also extracts by far the highest number of relations among all the systems that we considered in our evaluation (see Table 15.22).

```
Stanford Open IE:
(1) she;    was confirmed on;   August 6 2009
(2) she;    was confirmed;      becoming
(3) she;    becoming;           Supreme Court Justice of Hispanic descent
(4) she;    was confirmed;
(5) she;    becoming;           first Supreme Court Justice
(6) she;    becoming;           Supreme Court Justice
(7) she;    becoming;           first Supreme Court Justice of Hispanic
descent
```

Figure 15.6 Relations extracted by Stanford Open IE

OLLIE   As opposed to ClausIE and Stanford Open IE, which focus on restructuring the argument slot, OLLIE puts the emphasis on conveying a range of alternative relational phrases by including, excluding and rearranging phrasal components, resulting in comparatively long relational phrases (e.g., (2), (3) and (4) in Figure 15.7). By contrast, its argument phrases are generally relatively short (see also Table 15.22 and Table 15.26).

```
Ollie:
(1) she;               was confirmed on;           August 6, 2009
(2) He;                nominated Sonia Sotomayor on;  May 26
(3) He;                nominated Sonia Sotomayor;     2009
(4) He;                nominated 2009 on;           May 26
(5) Sonia Sotomayor;   be nominated 2009 on;        May 26
(6) He;                nominated 2009;              Sonia Sotomayor
(7) 2009;              be nominated Sonia Sotomayor on;  May 26
```

Figure 15.7 Relations extracted by OLLIE

REVERB   OLLIE's predecessor system REVERB lays the focus on preventing the extraction of incoherent and uninformative relations, thereby making very conservative choices that lead to a high precision of the extracted relational tuples, as the automatic comparative analysis in Section 15.3.1.1 has shown. However, its output often lacks in comprehensiveness, resulting in tuples that are commonly short and accurate, but miss a lot of information contained in the source sentences (see Figure 15.8).

```
ReVerb:
(1) He;        nominated;        Sonia Sotomayor
(2) she;       was confirmed on;      August 6, 2009
```

Figure 15.8 Relations extracted by REVERB

OpenIE-4 and OpenIE-5 Similar to REVERB, OpenIE-4 and OpenIE-5 commonly succeed in extracting highly accurate tuples that are composed of relational and argument phrases which represent minimal semantic units (e.g., (1) and (3) in Figure 15.9). However, as opposed to REVERB, they take a less conservative approach in the extraction process, allowing for a higher recall of the extracted relations, which is reflected in the higher coverage of input tokens in the output tuples (see Table 15.22 and Table 15.26).

```
OpenIE-4:
(1) he;        nominated;        Sonia Sotomayor
(2) she;       was confirmed;
(3) she;       was becoming;    the first Supreme Court Justice of Hispanic
descent
```

Figure 15.9 Relations extracted by OpenIE-4

MinIE MinIE was built on top of ClausIE with the goal of minimizing both relational and argument phrases by identifying and removing parts that are overly specific. Indeed, as Figure 15.10 shows, it succeeds in shortening the argument phrases to their core information (e.g., (1), (2), (5) and (6) in Figure 15.10), extracting a large number of correct tuples with atomic arguments that cover most of the information contained in the input. However, similar to OLLIE, MinIE tends to produce over-generated predicates that may impede the processing of the output in downstream applications.

```
MinIE:
(1) He;    nominated Sonia Sotomayor on;        May 26
(2) Ho;    nominated Sonia Sotomayor to replace;    David Souter
(3) He;    nominated;                           Sonia Sotomayor
(4) He;    be replace;                          David Souter
(5) she;   was confirmed on August 6 2009 becoming first Supreme Court
Justice of;                                     Hispanic descent
(6) she;   was confirmed becoming first Supreme Court Justice of;
                                                Hispanic descent
```

Figure 15.10 Relations extracted by MinIE

PropS PropS is one of the few Open IE systems under consideration that aim to extract n-ary relations with the objective of obtaining as complete an extraction as possible. It achieves to split the argument phrases into indivisible units, with each of them conveying a separate fact. Thus, PropS extracts highly accurate and complete facts from the source sentences (see Figure 15.11).

```
PropS:
(1) He;        nominated;    Sonia Sotomayor; on May 26, 2009; to replace
David Souter
(2) she;       confirmed;    on August 6, 2009; becoming the first Supreme
Court Justice of Hispanic descent
(3) ;          first;        the Supreme Court Justice of Hispanic descent
```

Figure 15.11 Relations extracted by PropS

RnnOIE Similar to PropS, RnnOIE targets the extraction of n-ary relations in order to capture facts as complete as possible from the input. In doing so, it follows a verb-centric approach that aims to extract a relation for every verbal predicate in the source sentence (see Figure 15.12).

```
RnnOIE:
(1) He;        nominated;    Sonia Sotomayor; on May 26, 2009; to replace
David Souter
(2) ;          replace;      David Souter
(3) she;       confirmed;    on August 6, 2009; becoming the first Supreme
Court Justice of Hispanic descent
(4) she;       becoming;     the first Supreme Court Justice of Hispanic
descent
```

Figure 15.12 Relations extracted by RnnOIE

15.3.2 Sentence Splitting as a Pre-processing Step

In the section below, we examine how discourse-aware sentence splitting affects the performance of state-of-the-art Open IE systems, both in terms of the coverage and accuracy of the extracted relational tuples, and the enrichment of the output with contextual information.

15.3.2.1 Automatic Comparative Analysis of the Outputs

After evaluating Graphene as a reference Open IE implementation, we examine the impact of our proposed discourse-aware TS approach on the accuracy and coverage of the relations extracted by a set of state-of-the-art Open IE systems. For this purpose, we conduct a quantitative comparison of the systems' recall and precision scores on the OIE2016 and CaRB benchmarks with and without using our sentence splitting framework as a pre-processing step, and analyze whether their particular deficiencies can be eliminated when operating on the structurally simplified sentences.

OIE2016 Figure 15.13 demonstrates the effectiveness of our proposed TS process that splits complex sentences into a set of minimal propositions with a more regular syntax, when being applied as a pre-processing step for the task of Open IE. It shows the PR curves of the Open IE baseline systems that have been executed on the OIE2016 benchmark as a stand-alone system (dashed lines) (for details, see Section 15.3.1.1), and within our TS-Open IE pipeline, where they act as the relation

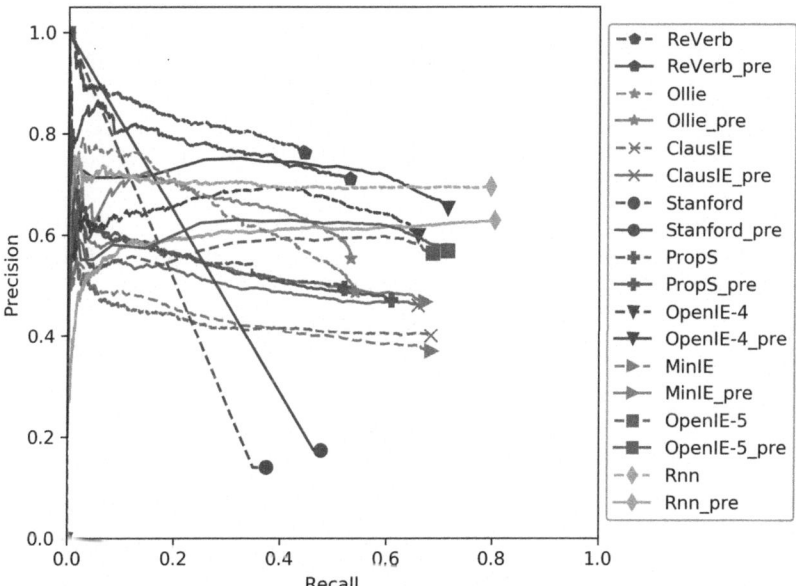

Figure 15.13 Comparative analysis of the performance of state-of-the-art Open IE systems on the OIE2016 benchmark *with* (solid lines) and *without* (dashed lines) sentence splitting as a pre-processing step

extraction component after the transformation stage (solid lines). The average precision, recall, F_1 and AUC scores when operating on top of the simplified sentences are listed in Table 15.29, while the corresponding improvements—as compared to their stand-alone counterparts—are given in Table 15.30.

The scores in Table 15.30 show that when splitting complex sentences into a set of syntactically simplified propositions, all systems under consideration gain in F_1 and AUC, except for OLLIE and RnnOIE. The highest improvement in AUC is achieved by REVERB, yielding a 39.19% increase over the output produced

Table 15.29 Average precision, recall, F_1 and AUC scores of a variety of state-of-the-art Open IE approaches on the OIE2016 benchmark when using our reference TS implementation DisSim as a pre-processing step

System	Precision	Recall	F_1	AUC
REVERB	**0.776**	0.531	0.631	0.412
OLLIE	0.667	0.533	0.593	0.353
ClausIE	0.503	0.660	0.571	0.332
Stanford Open IE	0.175	0.477	0.256	0.275
PropS	0.522	0.609	0.562	0.326
OpenIE-4	0.727	0.717	**0.722**	**0.519**
MinIE	0.533	0.674	0.595	0.359
OpenIE-5	0.610	0.717	0.659	0.435
RnnOIE	0.596	**0.806**	0.685	0.480

Table 15.30 Improvements in precision, recall, F_1 and AUC on the OIE2016 benchmark when using our reference TS implementation DisSim as a pre-processing step

System	Precision	Recall	F_1	AUC
REVERB	−7.18%	+19.06%	+8.42%	**+39.19%**
OLLIE	+1.06%	−1.48%	−0.34%	−1.12%
ClausIE	+16.98%	−3.51%	+8.14%	+12.93%
Stanford Open IE	**+25.00%**	**+27.20%**	**+25.49%**	+35.47%
PropS	−6.12%	+16.89%	+4.46%	+12.41%
OpenIE-4	+9.98%	+8.64%	+9.39%	+19.59%
MinIE	+23.67%	−1.75%	+12.48%	+21.28%
OpenIE-5	+4.99%	+4.22%	+4.60%	+9.02%
RnnOIE	−14.98%	+0.88%	−8.30%	−14.13%

when acting as a stand-alone system, followed by Stanford Open IE (+35.47%) and MinIE (+21.28%). On the contrary, OLLIE's (−1.12%) and RnnOIE's (−14.13%) AUC scores somewhat decline. Regarding F_1, the picture is similar. While OLLIE (−0.34%) and RnnOIE (−8.30%) lose in F_1, all other systems under consideration enhance their score, above all Stanford Open IE (+25.49%), MinIE (+12.48%) and OpenIE-4 (+9.39%). On closer examination, we find that some approaches mainly improve in precision, whereas others primarily profit from a boost in recall. The former include for example Stanford Open IE (+25.00%), MinIE (+23.67%) and ClausIE (+16.98%), while the latter comprises systems such as Stanford Open IE (+27.20%), REVERB (+19.06%) and PropS (16.89%) (for details, see Section 15.3.2.2).

Analogous to Table 15.22, Table 15.31 reports a number of descriptive statistics on the relations extracted from the sentences included in the OIE2016 benchmark. The only difference is that this time we used our proposed TS approach as a pre-processing step to split the source sentences into a set of syntactically simplified

Table 15.31 Descriptive statistics on the relations extracted from the 3,200 sentences of the OIE2016 corpus when using our reference TS implementation DISSIM as a pre-processing step. We report scores for a number of basic statistics, including the total number of relational tuples extracted from the input (#E); the average number of facts extracted per source sentence (#E/S); the percentage of non-redundant extractions (%$E_{non-redundant}$); the number of sentences from which not even one fact is extracted (#S<1E); the average length of the extracted relational tuples on the basis of the average number of tokens included in the phrases in subject, predicate and object position; and the percentage of tokens from the input that are contained in at least one of the extracted relational tuples (coverage)

System	#E	#E/S	%$E_{non-redundant}$	#S<1E	avg. triple length (s, p, o)	coverage
REVERB	5,142	1.61	98.74%	428	9.75 (2.15, 2.51, 5.09)	0.46
OLLIE	7,109	2.22	98.02%	334	9.23 (2.25, 2.80, 4.19)	0.35
ClausIE	11,647	3.64	94.14%	40	10.50 (2.44, 1.66, 6.40)	**0.71**
Stanford Open IE	**21,256**	**6.64**	85.78%	245	**7.46** (1.55, 1.93, 3.98)	0.36
PropS	10,464	3.27	**99.68%**	20	10.21 (2.87, 1.04, 6.30)	0.66
OpenIE-4	8,705	2.72	98.98%	71	10.87 (2.60, 1.79, 6.48)	0.57
MinIE	10,865	3.40	97.24%	83	8.25 (1.95, 2.67, 3.63)	0.53
OpenIE-5	9,970	3.12	98.55%	80	10.44 (2.54, 1.82, 6.08)	0.55
RnnOIE	9,881	3.09	99.02%	**10**	10.59 (2.45, 1.89, 6.25)	0.60
Average	10,559.89	3.30	96.68%	145.67	9.70 (2.31, 2.01, 5.38)	0.53

propositions before applying the Open IE baseline systems to extract the relational tuples contained in the input.

The figures show that when operating on the simplified sentences, the average number of tuples extracted from the source sentences slightly increases (+3.08%). In particular, it can be seen that those systems that tend to overproduce relational or argument phrases, such as OLLIE (−13.58%), ClausIE (−16.49%) or MinIE (−23.35%), tend to return less extractions, whereas the remaining approaches are likely to extract a higher number of tuples as compared to when running directly on the complex input sentences. On average, the number of facts extracted per source sentence marginally increases (+3.12%), while at the same time the number of sentences from which no tuple is extracted significantly drops (−20.69%). The average triple length, too, slightly decreases by one token. It is particularly noticeable that REVERB, which was shown to be very conservative in its extractions, tends to produce tuples that are longer by more than two tokens on average. In contrast, PropS returns relations that are significantly shorter (−38.75%), indicating that a comparatively high amount of information that is contained in its high-arity relations when using the original complex input sentences are dropped. In addition, the overall coverage rate declines by 19.70%, suggesting that when extracting relations from the pre-processed sentences some information included in the source is lost. Noteworthy in this respect is that REVERB is the only approach that achieves a higher coverage of the input (+9.52%), while OLLIE (−40.68%), OpenIE-4 (−25.97%) and PropS (−23.26%) suffer from a particularly large decline regarding the fraction of input tokens included in the extracted relational tuples.

Table 15.32 Distribution of the arity of the relations extracted from the 3,200 sentences of the OIE2016 dataset when using our reference TS implementation DISSIM as a pre-processing step

System	unary	binary	ternary	quaternary	≥ quaternary
REVERB	–	66%	23%	8%	4%
OLLIE	–	100%	–	–	–
ClausIE	5%	77%	13%	4%	1%
Stanford Open IE	–	100%	–	–	–
PropS	13%	50%	23%	9%	5%
OpenIE-4	12%	55%	24%	7%	2%
MinIE	5%	75%	13%	5%	2%
OpenIE-5	10%	58%	22%	7%	3%
RnnOIE	11%	58%	25%	5%	1%

Table 15.32 provides an overview of the arity of the extracted relations, when using our reference TS implementation DISSIM as a pre-processing step. The numbers show that in this case all Open IE systems under consideration output n-ary extractions ($n > 2$), aside from OLLIE and Stanford Open IE, which still solely extract binary relations. For the remaining systems, we observe that about a third of the extracted relational tuples consist of at least three arguments.

CaRB   To support our findings from the previous section, we replicated our analysis on the customized version of the CaRB benchmark. Using its evaluation methodology in combination with the ground truth relations from the OIE2016 benchmark (see Section 14.3.3.1), we investigated the impact of applying our proposed TS approach as a pre-processing step to split complex input sentences into a set of syntactically simplified propositions. Figure 15.14 illustrates the outcome.

As before, it depicts the PR curves of the Open IE baseline systems that have been executed on the CaRB benchmark both as a stand-alone system (dashed lines) (for details, see Section 15.3.1.1.1), and within our TS-Open IE pipeline, where they act as the relation extraction component after the transformation stage (solid lines). The systems' overall precision, recall, F_1 and AUC scores when operating on top

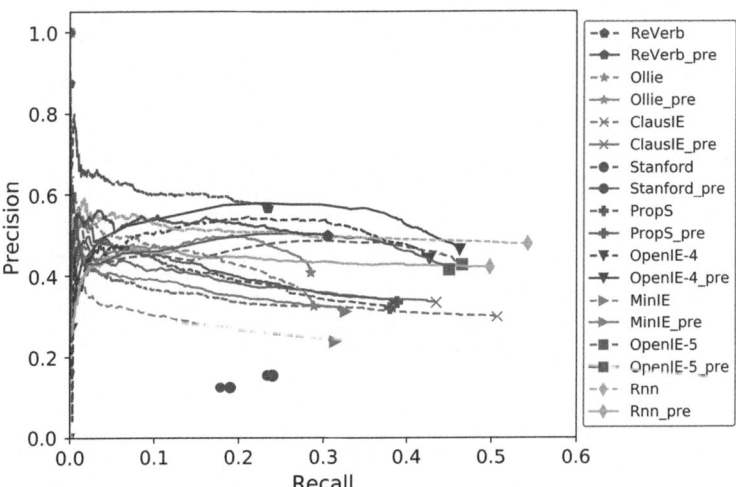

Figure 15.14 Comparative analysis of the performance of state-of-the-art Open IE systems on the customized CaRB benchmark *with* (solid lines) and *without* (dashed lines) sentence splitting as a pre-processing step

Table 15.33 Precision, recall, F_1 (at maximum F_1 point) and AUC of a variety of state-of-the-art Open IE approaches on the CaRB benchmark when applying our reference TS implementation DISSIM as a pre-processing step and using the gold tuples from the OIE2016 dataset in combination with the improved evaluation metrics from the CaRB framework

System	Precision	Recall	F_1	AUC
REVERB	**0.500**	0.305	0.379	0.160
OLLIE	0.432	0.281	0.341	0.134
ClausIE	0.336	0.433	0.379	0.168
Stanford Open IE	0.154	0.240	0.188	0.136
PropS	0.342	0.380	0.360	0.158
OpenIE-4	0.478	0.454	**0.466**	**0.250**
MinIE	0.323	0.316	0.319	0.120
OpenIE-5	0.426	0.443	0.434	0.215
RnnOIE	0.421	**0.498**	0.456	0.216

Table 15.34 Improvements in precision, recall, F_1 and AUC scores on the CaRB benchmark when applying our reference TS implementation DISSIM as a pre-processing step and using the gold tuples from the OIE2016 dataset in combination with the improved evaluation metrics from the CaRB framework

System	Precision	Recall	F_1	AUC
REVERB	−11.82%	**+29.79%**	+14.16%	+11.11%
OLLIE	+19.01%	+1.81%	+8.60%	+3.08%
ClausIE	+11.63%	−13.75%	+0.80%	−2.33%
Stanford Open IE	+23.20%	+26.32%	**+24.50%**	**+33.33%**
PropS	+5.56%	+0.53%	+3.15%	+3.27%
OpenIE-4	+3.02%	+9.13%	+6.15%	+16.82%
MinIE	**+31.84%**	+4.29%	+17.71%	**+33.33%**
OpenIE-5	−4.27%	−2.42%	−3.34%	−1.38%
RnnOIE	−11.92%	−8.29%	−10.24%	−21.17%

of the simplified sentences are listed in Table 15.33. The corresponding improvements achieved in comparison to their stand-alone counterparts are displayed in Table 15.34.

Though the order of the tested Open IE systems slightly changes with respect to each evaluation metric, the overall trend is similar to what could already be observed on the OIE2016 benchmark in the previous section. As shown in Table 15.34, all approaches under consideration gain both in F_1 and AUC, except for ClausIE (+0.80%, −2.33%), OpenIE-5 (−3.34%, −1.38%) and RnnOIE (−10.24%, −21.17%). What stands out is that, as before, Stanford Open IE (+24.50% and +33.33%) and MinIE (17.71% and 33.33%) are among the systems that profit most from simplifying the structure of the input sentences.

Moreover, in contrast to the results obtained on the OIE2016 benchmark, OLLIE now benefits from applying our TS approach as a pre-processing step with regard to all four metrics (+19.01%, +1.81%, +8.60%, +3.08%). This is likely due to the lexical matching function used in OIE2016 not penalizing extractions for misidentifying parts of an argument phrase in the relation slot, therefore already assigning high precision scores to the stand-alone version of OLLIE. CaRB's matching function, however, utilizes a tuple match that punishes such a misplacement. Since OLLIE is particularly prone to extracting overgenerated predicates (see Section 15.3.1.3), it greatly benefits from the split sentences, as they prevent it from producing overly long relational phrases, which leads to a sharp rise in precision by 19.01%. The same applies to MinIE, explaining its high boost in precision (+31.84%). Conversely, OpenIE-5 shows a slight decline in all four metrics (−4.27%, −2.42%, −3.34%, −1.38%) when running on the syntactically simplified sentences instead of the raw complex input data.

Furthermore, ClausIE shows a comparatively larger drop in recall (−13.74%), resulting in a marginally lower AUC score (−2.33%) as compared to when operating on the raw source sentences. This result can be explained by CaRB's scoring function. Although the CaRB framework aims for atomic relational tuples that represent a minimal semantic unit each, it uses a multi-match approach for measuring a system's recall in order to avoid penalizing it too severely if it puts information from multiple gold tuples into a single extraction. Therefore, CaRB's matching function is likely to yield a relatively high recall when operating on the original data, upon which ClausIE tends to produce overly long argument phrases through combining and rearranging phrasal elements of the input (see Section 15.3.1.3). When instead applying our TS approach as a pre-processing step, there is a risk that some of the information contained in the source sentences is lost throughout the transformation process, resulting in a decreased recall. Though, the split sentences greatly help in generating indivisible argument phrases.

Summary   The automatic comparative evaluation conducted on a set of state-of-
the-art Open IE baseline systems demonstrated the effectiveness of using our proposed discourse-aware TS approach as a pre-processing step. We showed that 78%
(OIE2016 benchmark) and 67% (CaRB benchmark) of the tested approaches benefit
(in terms of an improved F_1 and AUC score) from splitting complex sentences into
a set of minimal propositions with a more regular syntax, upon which the relation
extraction step is then performed.

15.3.2.2 Qualitative Analysis of the Extracted Relations

Figures 15.16 to 15.23 illustrate the outputs produced by the Open IE baselines on
the example sentence from Section 15.3.1.3, when using as input the set of syntactically simplified sentences from Figure 15.15, which were generated by our proposed
discourse-aware TS approach. For all Open IE systems, we see that a substantial proportion of the extracted relational tuples are enriched with contextual information.
In all cases, this supplementary information is correctly assigned to the extraction
to which it refers. In the following, we will discuss the effects of applying our TS
approach as a pre-processing step in more detail, together with its implications on
the systems' overall precision and recall scores.[16]

```
DisSim:
(1) #1    0    He nominated Sonia Sotomayor.
(1a)                S:PURPOSE    This was to replace David Souter.
(1b)                S:TEMPORAL   This was on May 26, 2009.

(2) #2    0    She was confirmed.
(2a)                S:TEMPORAL   This was on August 6, 2009.

(3) #3    0    She was becoming the first Supreme Court Justice of
Hispanic descent.
```

Figure 15.15 Simplified output generated by our reference TS implementation DisSim on
the following example sentence: *"He nominated Sonia Sotomayor on May 26, 2009 to replace
David Souter; she was confirmed on August 6, 2009, becoming the first Supreme Court Justice
of Hispanic descent."*

---

[16] We refer here to the precision and recall scores achieved on the OIE2016 benchmark.

ReVerb When acting as a stand-alone system, REVERB achieves a particularly high precision of the extracted relational tuples, though at the cost of a comparatively low recall rate. The reason for this is that REVERB tends to make very conservative choices in the extraction process, leading to the lowest number of output relations with the shortest average triple length (see Table 15.22). However, when using our TS approach as a pre-processing step, REVERB succeeds in extracting a larger number of relations from the input sentences with an increased average triple length, resulting in an improved recall score. Figure 15.16 supports these findings. It shows that REVERB now detects all the relations contained in the input, whereas before, when operating on the original sentence, it was only able to identify the two main relations of the source, without any additional contextual information (see Figure 15.8). Still, overall precision slightly decreases, most likely due to errors that were introduced during the transformation stage, leading to incorrectly split sentences.

```
ReVerb (with pre-processing):
(1) #1   0   He;     nominated;      Sonia Sotomayor
(1a)              S:PURPOSE    to replace David Souter
(1b)              S:TEMPORAL   on May 26 , 2009

(2) #2   0   she;   was confirmed;
(2a)              S:TEMPORAL   on August 6 , 2009

(3) #3   0   she;   was becoming;   the first Supreme Court Justice of
Hispanic descent
```

Figure 15.16 Relations extracted by REVERB when using our reference TS implementation DISSIM as a pre-processing step

Ollie Regarding recall and precision, OLLIE's performance does not change significantly whether or not applying our proposed TS approach as a pre-processing step. Using our reference implementation DISSIM, it slightly improves in precision, while losing somewhat in recall. However, as the output in Figure 15.17 illustrates, the simplified sentence structure helps in solving OLLIE's main problem: overgenerated predicates. When operating on the split sentences, the number of quasi-redundant extractions which stem from OLLIE's focus on performing a great many of phrasal permutations for generating the relational phrases of its extractions, is dramatically reduced. Instead of rearranging phrasal components in the relational phrases in various ways (see Figure 15.7), they are now appended to the extracted tuples in the form of additional contextual information.

```
Ollie (with pre-processing):
(1) #1    0    he;      nominated;      Sonia Sotomayor
(1a)                S:PURPOSE   to replace David Souter
(1b)                S:TEMPORAL  on May 26 , 2009


(2) #2    0    she;    was becoming;   the first Supreme Court Justice of
Hispanic descent
```

Figure 15.17 Relations extracted by OLLIE when using our reference TS implementation DISSIM as a pre-processing step

ClausIE ClausIE primarily benefits from a push in precision, while its recall slightly drops. As we see in Figure 15.18, this is no surprise. The syntactically simplified sentences prevent ClausIE from generating argument phrases by simply combining phrasal elements from the input in various different orders. In that way, its main problem—overly long argument phrases that subsume other self-contained facts (see Figure 15.5)—is resolved. Rather, ClausIE is likely to output arguments that represent an indivisible component in terms of a minimal semantic unit when acting on the split sentences provided by our TS approach.

```
ClausIE (with pre-processing):
(1) #1    0    he;      nominated;    Sonia Sotomayor
(1a)                S:PURPOSE    to replace David Souter.
(1b)                S:TEMPORAL   on May 26, 2009.

(2) #2    0    she;    was confirmed;
(2a)                S:TEMPORAL   on August 6, 2009.

(3) #3    0    she;    was becoming;   the first Supreme Court Justice of
Hispanic descent
```

Figure 15.18 Relations extracted by ClausIE when using our reference TS implementation DISSIM as a pre-processing step

Stanford Open IE Of all tested systems, Stanford Open IE shows the highest increase in both precision and recall. While it still tends to overproduce argument phrases by adding, deleting and restructuring optional phrasal complements— thereby generating a large number of redundant extractions—, it profits from the simplified sentences by enriching its extractions with contextual information (see e.g., (2) and (2a) in Figure 15.19), leading to a higher recall of its output. As Figure 15.19 shows, Stanford Open IE is now even able to detect relations that have not been identified before (see Figure 15.6) when operating on the raw input sentences ((1), (1a) and (1b) in Figure 15.19). At the same time, the inaccurate extraction ⟨*she; was confirmed; becoming*⟩ is eliminated from the output.

```
Stanford Open IE (with pre-processing):
(1) #1   0   he;     nominated;        Sonia Sotomayor
(1a)                 S:PURPOSE    to replace David Souter
(1b)                 S:TEMPORAL   on May 26 , 2009


(2) #2   0   she;    was;              confirmed
(2a)                 S:TEMPORAL   on August 6 , 2009


(3) #3   0   she;    was becoming;   first Supreme Court Justice


(4) #4   0   she;    was becoming;   first Supreme Court Justice of
Hispanic descent


(5) #5   0   she;    was becoming;   Supreme Court Justice


(6) #6   0   she;    was becoming;   Supreme Court Justice of Hispanic
descent
```

Figure 15.19 Relations extracted by Stanford Open IE when using our reference TS implementation DISSIM as a pre-processing step

PropS PropS profits from a clear boost in recall when operating upon the structurally simplified sentences. This can be explained by PropS's tendency to incorporate relations in the argument slot of a superordinate extraction (e.g., ⟨*she becoming the first Supreme Court Justice of Hispanic descent*⟩ in (2) in Figure 15.11) when acting on the raw input sentences. In contrast, the split sentences encourage the extraction of a separate relational tuple for such cases (see (3) in Figure 15.20). Moreover, while PropS is likely to produce a large number of n-ary relations when given the original complex sentences as input, it generates an output where additional arguments are preferably expressed in contextual arguments that are attached to the respective parent relation (e.g., (1a) and (1b) in Figure 15.20) when applying our TS approach as a pre-processing step.

```
PropS (with pre-processing):
(1) #1    0    He;     nominated;       Sonia Sotomayor
(1a)                   S:PURPOSE     replace David Souter
(1b)                   S:TEMPORAL    on May 26 , 2009


(2) #2    0    she;    confirmed;
(2a)                   S:TEMPORAL    on August 6 , 2009


(3) #3    0    she;    becoming;    the first Supreme Court Justice of
Hispanic descent
```

Figure 15.20 Relations extracted by PropS when using our reference TS implementation DisSim as a pre-processing step

OpenIE-4 and OpenIE-5 OpenIE-4 and OpenIE-5 both benefit from the syntactically simplified sentences in terms of recall and precision. Although they already achieve a relatively high recall rate when operating on the original complex input sentences, they still tend to miss some of the contextual information contained in them (see Figure 15.9), which they are able to detect when given the simplified sentences as input (e.g., (1a) and (1b) in Figure 15.21), leading to an increased recall score.

```
OpenIE-4 (with pre-processing):
(1) #1    0    he;     nominated;   Sonia Sotomayor
(1a)                   S:PURPOSE     to replace David Souter.
(1b)                   S:TEMPORAL    on May 26, 2009.

(2) #2    0    she;    was;    confirmed
(2a)                   S:TEMPORAL    on August 6, 2009.

(3) #3    0    she;    was becoming;   the first Supreme Court Justice of
Hispanic descent
```

Figure 15.21 Relations extracted by OpenIE-4 when using our reference TS implementation DisSim as a pre-processing step

MinIE When using our TS approach as a pre-processing step, MinIE achieves a notable growth in precision, while losing somewhat in recall. This is due to the fact that, similar to Ollie, when operating on the original data, it is prone to producing overgenerated predicates (see Figure 15.10), a problem that vanishes when taking the simplified sentences as input. Rather than incorporating and permuting phrasal complements in the relational phrases, MinIE then adds them to the extracted tuples as supplementary contextual information, resulting in a set of extractions that is much smaller, but at the same time much more precise (see Figure 15.22).

```
MinIE (with pre-processing):
(1) #1   0   He;     nominated;     Sonia Sotomayor
(1a)               S:PURPOSE    to replace David Souter
(1b)               S:TEMPORAL   on May 26 , 2009

(2) #2   0   she;    was confirmed;
(2a)               S:TEMPORAL   on August 6 , 2009

(3) #3   0   she;    was becoming the first Supreme Court Justice of;
Hispanic descent
```

Figure 15.22 Relations extracted by MinIE when using our reference TS implementation DISSIM as a pre-processing step

RnnOIE   RnnOIE is the only Open IE system under consideration that is clearly adversely affected by applying our TS approach as a pre-processing step. Its precision considerably drops by almost 15%, whereas there is a slight increase by less than 1% in recall. This might be due to the fact that RnnOIE was explicitly trained on the OIE2016 dataset and is therefore tailored to its specific features. Accordingly, the splitting errors introduced by our reference TS implementation DISSIM have a particularly strong negative impact on its performance, leading to a sharp decrease in the precision of its extracted tuples. Figure 15.23 illustrates the output produced by RnnOIE when operating on the simplified sentences.

```
RnnOIE (with pre-processing):
(1) #1   0   He;     nominated;     Sonia Sotomayor
(1a)               S:PURPOSE    to replace David Souter
(1b)               S:TEMPORAL   on May 26 , 2009

(2) #2   0   she;    confirmed;
(2a)               S:TEMPORAL   on August 6 , 2009

(3) #3   0   she;    becoming;   the first Supreme Court Justice of
Hispanic descent
```

Figure 15.23 Relations extracted by RnnOIE when using our reference TS implementation DISSIM as a pre-processing step

In general, we note that in many cases, the specific characteristics of the Open IE systems with regard to the type and boundaries of both the relational and argument phrases dissipate when operating on the simplified sentences. For instance, the overgenerated predicates that can be frequently observed in the output of OLLIE and MinIE disappear, as well as ClausIE's overly long argument phrases. Only Stanford Open IE still shows a strong tendency towards overproducing argument phrases. Hence, we conclude that by applying our discourse-aware TS approach as a pre-processing step, our TS-Open IE pipeline is capable of correcting erroneous extractions, as indicated by the improved overall precision. Moreover, the higher recall scores indicate that the split sentences substantially facilitate the extraction of the relations contained in the input.

15.3.2.3 Analysis of the Lightweight Semantic Representation of Relational Tuples

In Chapter 12, we presented a novel formal semantic representation for Open IE that allows for a *canonical context-preserving representation of relational tuples*. In the following, we will briefly examine its main properties in the light of the experiments outlined above.

Simplistic Canonical Predicate-Argument Structure   The minimal propositions generated by our TS approach not only reduce the complexity of the relation extraction step in the Open IE task, but also lead to a canonical predicate-argument structure of the extracted relational tuples.

Reduced complexity of the relation extraction step   The formal semantic representation for relational tuples that we propose is primarily aimed at complex input sentences. Using our TS approach from Chapter II, they are first split into a set of minimal semantic units, where each proposition represents a separate fact. In that way, complex polypredicative constructions are transformed into mono-predicative units and trimmed down to their essential constituents, i.e. elements that are also part of their clause types. In addition, a specified set of phrasal units is extracted (see Section 4.1). Thus, long nested structures are broken up and transformed into a normalized subject-verb-object structure (or simple variants thereof, see Table 4.1). Consequently, the 1-to-n problem when extracting predicate-argument tuples from complex sentences is scaled down to a *1-to-1 predicate mapping problem*, where exactly one relational tuple is generated from each simplified sentence. Thus, the complexity of the relation extraction step is reduced. Beyond that, shortening each simplified sentence to its core components further minimizes the complexity of this task, ensuring that overly specific argument phrases that subsume other self-contained facts are prevented from being generated.

According to the human analyses presented in Section 15.1.2.1, we succeed in splitting complex source sentences into a fine-grained output of simplified sentences that are to a large extent syntactically well-formed and preserve the meaning of the input. In particular, we achieve an average score of 1.45 for the "simplicity" dimension on the three TS corpora. This result is close to the maximum score of 2, which indicates that a given output sentence presents an atomic semantic unit. Accordingly, we conclude that our approach largely succeeds in decomposing complex input sentences into *minimal propositions with a simplistic canonical structure* that can be leveraged by downstream Open IE approaches.

As demonstrated by the empirical findings described in Section 15.3.2.1, the fine-grained representation of sentences in the form of atomic semantic units is easier to analyze for state-of-the-art Open IE approaches, resulting in a higher precision and recall of the extracted relational tuples with regard to six out of the nine tested systems (see Tables 15.30 and 15.34). Consequently, when operating on the simplified sentences, Open IE frameworks are more likely to return correct relational tuples and less prone to miss information from the source sentence in the extracted tuples.

Canonical predicate-argument structure   The minimal propositions that present a simplistic canonical structure inherently lead to a normalized representation of the relational tuples extracted by an Open IE system. Each simplified sentence results in a *canonical (mostly) binary*[17] *predicate-argument structure* that is enriched with contextual information in order to avoid producing an isolated set of relational tuples that are hard to interpret (for details, see below).

When operating on the original complex input sentences, the characteristics of the output extracted by state-of-the-art Open IE systems vary widely. As outlined in the manual analysis in Section 15.3.1.3, there is no standardized output scheme among the different approaches. While some of the frameworks are limited to

---

[17] For better clarity, we appended relations that result from simple contexts as additional arguments to their respective parent tuples, resulting in n-ary relations with $n > 2$ in Table 15.32, as illustrated in the following example:

- core tuple: ⟨*Albert Einstein; won; the Nobel Prize*⟩
- context tuple: ⟨*This; was; in 1921*⟩

Core and context tuple are merged into the following ternary tuple. ⟨*Albert Einstein; won; the Nobel Prize, in 1921*⟩. Note that in their original representation, both tuples represent binary relations, though. Thus, according to the findings presented in Table 15.32, the overwhelming majority of the extractions represent binary predicate-argument structures, while only 7% of the extracted relational tuples are unary.

the extraction of binary relations, others output n-ary relations (typically without establishing a semantic connection between associated tuples). Moreover, many Open IE systems tend to produce a large number of quasi-redundant extractions. This is due to the fact that they are prone to overgenerate predicate and argument phrases by adding, deleting and restructuring phrasal complements. The resulting output often not only is redundant but also presents overly specific predicates, as well as overly long argument units, generally hurting the performance of downstream applications Gashteovski et al. (2017).

As the analysis in Section 15.3.2.2 reveals, the semantic representation that we propose normalizes the extracted relational tuples by *reducing the predicate and argument slots to their essential components*, while *additional information can be found in associated context tuples*. In that way, overgenerated predicate and argument slots, as well as (quasi-)redundant extractions are prevented for the most part, resulting in a simplistic unified canonical representation of relational tuples.

Contextual Information Previous work in the area of Open IE has mainly focused on the extraction of isolated relational tuples, resulting in a loose arrangement of tuples that lack important contextual information. To preserve the coherence of the extracted predicate-argument structures, the semantic representation that we propose extends the shallow semantic representation of state-of-the-art Open IE approaches by incorporating their semantic context. For this purpose, it captures intra-sentential rhetorical structures and hierarchical relationships between the relational tuples.

Contextual hierarchy To avoid producing a set of disconnected extractions, the proposed semantic representation specifies a hierarchical order between the extracted relations. For this purpose, it distinguishes *different levels of context*. Each tuple is assigned to a contextual layer – the lower the allotted layer, the more relevant is the information contained in it. In that way, related tuples are linked to one another, thus establishing a contextual hierarchy between them. According to the results presented in Section 15.2.1.1, our approach succeeds in assigning the correct hierarchical relationship between a pair of propositions in almost 90% of the cases.

The lower part of Figure 15.24 illustrates an example. Tuples (6) and (8) are assigned to contextual layer 0, i.e. the lowest level of context. Accordingly, they depict the key information of the source sentence. Tuples (7), (9), (10) and (11), in contrast, are each allocated to context level 1, which means that they provide additional information about their respective parent tuple. This type of information can easily be exploited for example in downstream text summarization applications. In order to reduce the content of a given document to its core information, such approaches need only focus on the relations of the lowest contextual layers, containing the main information of the input.

Semantically typed relational tuples In addition to setting up a contextual hierarchy between the extracted relational tuples, the proposed semantic representation for Open IE supports the extraction of further meta information that determines how the tuples are semantically interconnected. For this purpose, it applies a selected set of rhetorical relations that link associated tuples. In that way, it

```
OpenIE-4's default representation:
(1) A fluoroscopic study; known;    as an upper gastrointestinal series
(2) the usage of barium;  lead;    to increased post operative
complications
(3) volvulus;             is suspected;
(4) non water soluble contrast; is;   mandatory
(5) the usage of barium;  can impede; surgical revision


OpenIE-4's output when using our novel semantic representation:
(6) #1   0   A fluoroscopic study;   is typically;  the next step in
management
(6a)           L:ELABORATION  #2
(6b)           L:CONTRAST     #3


(7) #2   1   This fluoroscopic study; is known;     as an upper
gastrointestinal series


(8) #3   0   non water soluble contrast; is;        mandatory
(8a)           L:CONTRAST     #1
(8b)           L:CONDITION    #6
(8c)           L:BACKGROUND   #4
(8d)           L:BACKGROUND   #5


(9) #4   1   The usage of barium; can impede;  surgical revision
(9a)           L:LIST         #5


(10) #5  1   The usage of barium; can lead;    to increased post
operative complications
(10a)          L:LIST         #4


(11) #6  1   Volvulus;                is suspected;
```

Figure 15.24 Comparison of the output generated by OpenIE-4 when using its *default representation* and when using our *proposed formal semantic representation*

allows for the representation of semantically typed relational tuples. Thus, the *semantic context* of the extracted tuples is preserved. The comparative analysis with the annotations contained in the RST-DT in Section 15.2.1.2 shows that we reach an average precision of 70% for the classification of the rhetorical relations that hold between a pair of propositions.

As depicted in Figure 15.24, OpenIE-4's default representation returns a disconnected set of predicate-argument structures. Our novel representation, in contrast, enriches the output with contextual information in terms of rhetorical relations. In that way, the extracted relations are put into a logical structure in the form of semantically typed relational tuples that preserve the semantic context of the extractions, resulting in a *more informative and coherent output* which is easier to interpret.

15.4   Sentence Splitting Corpus

In the following, we report the results of both the automatic evaluation and the manual analysis performed on the sample of pairs of aligned complex source and simplified target propositions from our sentence splitting corpus MINWIKISPLIT.

15.4.1 Automatic Metrics

The results from the automatic evaluation, carried out on a random sample of 1,000 sentence pairs from MINWIKISPLIT, are provided in Table 15.35.

Table 15.35 Results of the automatic evaluation procedure on a random sample of 1,000 sentences from MINWIKISPLIT. We report scores for a number of descriptive statistics, including the average sentence length of the simplified target sentences in terms of the average number of tokens per output sentence (#T/S) and the average number of simplified output sentences per complex input (#S/C). Furthermore, in order to assess conservatism, we measure the percentage of sentences that are copied from the source without performing any simplification operation (%SAME) and the averaged word-based Levenshtein distance from the input (LD_{SC}). In addition, we report the average SAMSA and $SAMSA_{abl}$ scores of both the complex source and the simplified target sentences in our sample from the MINWIKISPLIT corpus

	#T/S	#S/C	%SAME	LD_{SC}	SAMSA	$SAMSA_{abl}$
Complex source	30.75	1.18	100	0.00	0.36	0.94
Simplified target	12.12	3.84	0.00	17.73	0.40	0.48

The figures demonstrate that on average our proposed sentence splitting corpus contains four simplified target sentences per complex source sentence, with every target proposition consisting of 12 tokens. Moreover, no input is simply copied to the output, but rather split into at least two smaller components. Both the high averaged Levenshtein distance of almost 18 and the SAMSA score (0.40) confirm previous findings. According to Sulem et al. (2018b), the latter is highly correlated with structural simplicity and grammaticality, indicating that the output sentences contained in our corpus are grammatically sound and present a simpler syntax than the input. With 0.48, we reach a decent score for the simplified target sentences with regard to SAMSA$_{abl}$, too, which has a high correlation with meaning preservation Sulem et al. (2018b).

15.4.2 Manual Analysis

The results of the human evaluation are displayed in Table 15.36. The inter-annotator agreement was computed using Cohen's quadratic weighted κ Cohen (1968). The obtained rates were 0.24, 0.25 and 0.75 for grammaticality, meaning preservation and structural simplicity, respectively, indicating a fair to substantial agreement between the annotators. System scores were calculated by averaging over the annotators' scores and the 300 randomly sampled sentences.

Table 15.36 Averaged human evaluation ratings on a random sample of 300 complex-simple sentence pairs from MINWIKISPLIT. Grammaticality, meaning preservation and structural simplicity are measured using a 1 (very bad) to 5 (very good) scale. A −2 to 2 scale is used for scoring the *structural* simplicity of the output relative to the input sentence

Grammaticality	4.36
Meaning preservation	4.10
Structural simplicity	1.43

The scores show that we succeed in producing simplified target sequences that reach a high level of grammatical correctness and almost always perfectly preserve the original meaning of the input. The third dimension under consideration, structural simplicity, which captures the degree of minimality in the simplified sentences, scores high values, too. However, we observed some room for improvement here. Consequently, in the future, we plan to implement stricter heuristics for sorting out output sequences that still mix multiple semantically unrelated propositions in order to ensure that each simplified target proposition represents an atomic semantic unit.

Part V
Conclusion

Summary and Contributions

16

This thesis contributes to the areas of TS and Open IE by critically analyzing existing approaches and proposing a new discourse-aware TS approach, as well as an Open IE framework that leverages its simplified output. In this section, we look back on the main contributions of each chapter.

16.1 Chapter I

Having defined the problem and the associated research questions and hypotheses to be addressed in this work (Chapter 1), **Chapter I** presents two comprehensive surveys of the state of the art in TS (Section 2.1) and Open IE (Section 2.3). In this context, six important research gaps are identified:

(TS-1) State-of-the-art TS approaches largely ignore the task of sentence splitting.

(TS-2) In the field of TS, there is little research on dealing with discourse-related issues caused by the rewriting transformations.

(TS-3) Current syntactic TS approaches follow a very conservative approach in that they tend to retain the input rather than transforming it.

(Open IE-1) Relations often span over long nested structures or are presented in a non-canonical form that cannot be easily captured by a small set of extraction patterns. Therefore, such relations are commonly missed by state-of-the-art Open IE approaches.

(Open IE-2) Current Open IE systems tend to extract relational tuples with long argument phrases that mix multiple, potentially semantically unrelated propositions, posing a challenge for downstream applications.

© The Author(s), under exclusive license to Springer Fachmedien Wiesbaden GmbH, part of Springer Nature 2022
C. Niklaus, *From Complex Sentences to a Formal Semantic Representation using Syntactic Text Simplification and Open Information Extraction*,
https://doi.org/10.1007/978-3-658-38697-9_16

(Open IE-3) Most state-of-the-art Open IE approaches lack the expressiveness needed to properly represent complex assertions, resulting in incomplete, uninformative or incoherent extractions that have no meaningful interpretation or miss critical information asserted in the input sentence.

Associated to this chapter is the publication Niklaus et al. (2018).

16.2 Chapter II

Based on the research gaps identified in Section 1.3, **Chapter II** presents a discourse-aware TS approach that focuses on the task of sentence splitting and thus the type of rewriting operation that has been ignored to a large extent in the TS research so far **(TS-1)**. The goal of our proposed framework is to break down sentences with a complex syntax into a set of *minimal propositions*, i.e. a sequence of sound, self-contained utterances, with each of them presenting a single event that cannot be further decomposed into meaningful propositions. For this purpose, we manually define a small set of 35 linguistically principled transformation rules that decompose clausal and phrasal elements from a given source sentence and turn them into stand-alone sentences that present a simplified structure. In that way, we aim to overcome the conservatism exhibited by state-of-the-art syntactic TS approaches, i.e. their tendency to retain the input rather than transforming it into a simplified output **(TS-3)**.

However, when splitting complex input sentences into a sequence of self-contained propositions without preserving their semantic context, our approach is prone to end up with a set of incoherent utterances which lack important contextual information that is needed for a proper interpretation of the output. Hence, such isolated propositions easily mislead to incorrect reasoning in downstream Open IE applications. To address this challenge, we develop a novel contextual representation that puts a *semantic layer on top of the simplified sentences* by establishing a semantic hierarchy between them **(TS-2)**. For this purpose, we first distinguish between sentence parts that provide key information (*"core sentences"*) and those that contribute only some piece of background information (*"contextual sentences"*). In this manner, we set up a contextual hierarchy between the decomposed utterances. In a second step, we then identify the semantic relationship that holds between a pair of split sentences, using a cue phrase mapping.

In that way, we generate a fine-grained representation of complex assertions in the form of hierarchically ordered and semantically interconnected sentences. The simplified propositions present a simple and more regular structure which is easier to process for downstream Open IE applications.

This framework is evaluated within our reference TS implementation DISSIM (see Chapter IV). Associated publications to this chapter are Niklaus et al. (2019b,a) and Niklaus et al. (2019c).

16.3   Chapter III

In **Chapter III**, we take advantage of the semantic hierarchy of simplified sentences generated by our discourse-aware TS approach, using it as a pre-processing step for *extracting semantically typed relational tuples* from complex assertions. We present a method that allows state-of-the-art Open IE systems to leverage the fine-grained representation of complex sentences in the form of hierarchically ordered and semantically interconnected propositions for facilitating the extraction of relational tuples **(Open IE-1** and **Open IE-2)** and to enrich their output with semantic information **(Open IE-3)** (Section 11.1), based on the following two hypotheses:

(1) We assume that shorter sentences with a more regular structure are easier to process and analyze for downstream Open IE applications. Hence, we expect to improve the performance of state-of-the-art Open IE systems in terms of the *precision* and *recall* of the extracted relational tuples when operating on the simplified sentences instead of dealing with the raw complex source sentences.

(2) We hypothesize that the semantic hierarchy generated by our discourse-aware TS approach can be leveraged to extract relations within the semantic context in which they occur. By enriching the output of state-of-the-art Open IE systems with additional meta information, we intend to create a novel *context-preserving representation* for the task of Open IE (Chapter 12). It aims to extend the shallow semantic representation of current approaches by capturing intra-sentential rhetorical structures and hierarchical relationships between the extracted relational tuples, thus producing an output that preserves important contextual information.

In addition, we present an Open IE reference system, Graphene, which implements a relation extraction pattern upon the semantic hierarchy of minimal propositions (Section 11.2).

The publications associated to this chapter are Cetto et al. (2018b), Niklaus et al. (2016) and Cetto et al. (2018a).

16.4   Chapter IV

In **Chapter IV**, we analyze the performance of our proposed discourse-aware TS approach with regard to the following two subtasks:

(1) splitting complex assertions into a set of minimal propositions (Section 15.1), and
(2) establishing a semantic hierarchy between the decomposed utterances (Section 15.2).

To assess the former, we conduct both thorough manual analyses and automatic evaluations across three widely used TS corpora. The results provide sufficient evidence that our TS approach succeeds in transforming sentences that present a complex syntax into a sequence of simplified sentences that are grammatically sound, represent atomic semantic units and have the same meaning as the corresponding original sentences. To examine the second subtask, we carry out an automatic comparative evaluation on the RST-DT dataset, which is complemented by an extensive manual analysis. Again, the experiments provide sufficient corroboration evidence to support the hypothesis that our framework is able to set up a semantic hierarchy between the split sentences in the form of hierarchically ordered and semantically interconnected sentences.

In addition, we assess the merits of our TS approach in augmenting the coverage and accuracy of the relational tuples extracted from complex input sentences in downstream Open IE applications (Section 15.3.2). The evaluation results demonstrate that the majority of Open IE approaches benefits in terms of AUC and F_1 when leveraging the fine-grained representation of complex assertions in the form of minimal propositions that present a canonical structure. Furthermore, we evaluate the performance of our reference Open IE implementation Graphene through a comparative analysis with a number of state-of-the-art Open IE baseline systems (Section 15.3.1). The results reveal that with an average precision of 78.5%, it succeeds in extracting highly accurate tuples at a high recall rate of 63.8%.

Finally, we outline to what extent the semantic hierarchy of minimal propositions generated by our discourse-aware TS approach can be leveraged to transform the shallow semantic representation of state-of-the-art Open IE systems into a novel canonical context-preserving representation of relational tuples.

Associated publications to this chapter are Niklaus et al. (2019b) and Cetto et al. (2018b).

16.5   Chapter V

In addition to summarizing the main contributions of each chapter, **Chapter V** revisits the research questions and associated hypotheses raised in the introductory part of this work (see Section 1.4). Under the perspective of the evaluation from the previous chapter, we draw the relevant conclusions (Chapter 17). Finally, we propose future research directions (Chapter 18).

Discussion and Conclusions 17

We analysed the performance of our proposed discourse-aware TS approach both intrinsically (see Section 17.1) and extrinsically (see Section 17.2), focusing on two research questions from which we derived seven hypotheses in total. In the following, we will revisit them and draw the relevant conclusions, on the basis of the experiments carried out and associated results from Chapter IV.

17.1 Semantic Hierarchy of Minimal Propositions

Research Question 1:

Is it possible to transform sentences that present a complex linguistic structure into a semantic hierarchy of minimal, self-contained propositions?

17.1.1 Subtask 1: Splitting into Minimal Propositions

The main goal of our TS approach is to transform sentences that present a complex syntax into a set of simplified sentences with a canonical structure, while preserving the meaning of the input. Accordingly, the transformation process that we propose must ensure that the resulting output sentences have the following three properties:

(i) *syntactic correctness,*
(ii) *minimality* and
(iii) *semantic correctness.*

© The Author(s), under exclusive license to Springer Fachmedien Wiesbaden GmbH, part of Springer Nature 2022
C. Niklaus, *From Complex Sentences to a Formal Semantic Representation using Syntactic Text Simplification and Open Information Extraction,*
https://doi.org/10.1007/978-3-658-38697-9_17

Below, we will discuss the performance of our TS approach with regard to the above criteria under the perspective of the evaluations that we performed.

17.1.1.1 Hypothesis 1.1: Syntactic Correctness

A complex sentence C can be transformed into a sequence of simplified sentences $T_1, \ldots, T_n, n >= 2$, such that each simple sentence T_i is grammatically sound.

Experimental support To assess the syntactic correctness of the generated simplified sentences, we performed both a thorough manual analysis and an automatic evaluation across three TS datasets from two different domains, Wikipedia articles and newswire texts. In a comparative analysis, we examined the output produced by our reference TS implementation DISSIM in contrast to the simplifications generated by several state-of-the-art TS baseline systems that have a strong focus on syntactic transformations through explicitly modelling splitting operations. In this context, we calculated the SAMSA scores of the simplifications produced by the TS systems under consideration. SAMSA is a recently proposed metric targeted at automatically measuring the syntactic complexity of sentences that was shown to highly correlate with human judgments on grammaticality. Moreover, to investigate the quality of the simplified output in more depth, we performed a manual analysis where pairs of complex source and simplified output sentences were rated by two in-house non-native, but fluent English speakers according to the criterion of grammaticality, using a Likert scale that ranged from 1 to 5. The detailed guidelines can be found in Table 14.1. To gain insight in what aspects of the simplification process are the most challenging for our proposed approach, we also performed a manual error analysis, thus figuring out the main sources of errors that led to simplified sentences that were not grammatically sound. Finally, we examined the accuracy of the specified transformation rules in terms of the percentage of input sentences that were correctly split into syntactically simplified output sentences.

Interpretation The comparative analysis we conducted revealed that our TS approach outperformed the state of the art in structural TS in generating simplified sentences that were grammatically sound. The framework we propose achieved the highest scores on all three TS corpora on the SAMSA metric (0.67, 0.57, 0.54), indicating that the output we produced presented a high level of syntactic correctness. This result was further supported by the human annotation ratings on the grammaticality dimension, where our reference implementation DISSIM (4.36) followed closely behind the YATS (4.40) and RegenT (4.64) systems. In fact, these two approaches were shown to act rather conservatively, performing only a small number of simplification operations (see Section 17.1.1.2). Therefore, they were much less prone to produce grammatically incorrect output, though at the cost of making less changes and ending up with output sentences that were still relatively complex. In contrast, our proposed TS approach was much less reluctant in splitting the input (see Section 17.1.1.2), while still generating simplifications that achieved a comparatively high level of syntactic correctness, as measured by the grammaticality score in the human evaluation. Moreover, the results of the manual analysis that we performed showed that, on average, correct simplification transformations were carried out in 81.1% of the cases, resulting in output sentences that were grammatically sound. Not each type of error that we identified to commonly occur during the simplification process of our framework (see Section 15.1.2.2) directly affected the syntactic correctness of the output. Still, three of them had an impact on the grammaticality of the generated simplified sentences, including morphological errors (16%), wrong split points (7%) and wrong order of the syntactic elements in the output (2%). However, these error classes only made up for about a quarter of the erroneous simplifications we found in our sample.

Conclusion Thus, to sum up, above findings suggest that our TS approach succeeds in splitting complex input sentences into simplified propositions that are to a large extent grammatically sound.

17.1.1.2 Hypothesis 1.2: Minimality

A complex sentence C can be transformed into a sequence of simplified sentences $T_1, \ldots, T_n, n >= 2$, such that each simple sentence T_i presents a minimal semantic unit, i.e. it cannot be further decomposed into meaningful propositions.

Experimental support To gauge the minimality property of the simplified sentences generated by our TS approach, we again calculated a number of automatic metrics and carried out a human evaluation. In these studies, the same set of TS corpora and baseline systems were used as before. In a first step, we determined a number of descriptive statistics that allowed conclusions about the simplicity of the output sentences. These included the average length of the simplified sentences, the average number of simplified output sentences per complex input, the percentage of sentences that were copied verbatim from the source without performing any simplification operation and the averaged word-based Levenshtein distance from the source. These scores provided evidence for how reluctant the underlying system was in splitting the input into simplified sentences. Next, we made use of the SAMSA score again, as it was shown to highly correlate with human judgments on structural simplicity, too. Moreover, analogous to the human evaluation from the previous section, we had two judges rate pairs of complex source and simplified output sentences based on their structural simplicity. In addition, we examined the frequency distribution of rule triggerings for each syntactic construct that is addressed by our TS approach, providing insights regarding the coverage of the specified rules with respect to the syntactic phenomena involved in complex sentence constructions. This problem was investigated more closely by manually checking the overall ratio of the simplification rules being fired whenever the corresponding clausal or phrasal construct was present in a given input sentence. Finally, the error analysis we conducted yielded further information on the extent to which our TS framework succeeded in splitting complex input sentences into minimal propositions.

Interpretation According to our experiments, the TS approach that we propose tends to extensively split the source sentences into simplified propositions. The descriptive statistics we calculated demonstrated that our reference implementation DISSIM reached the highest splitting rate (100%) among the TS systems under consideration. In addition, it not only returned by a large margin the highest number of structurally simplified sentences per source (2.82), but also the shortest sentences of all tested systems (11.01 tokens). These findings were confirmed by a comparatively high word-based Levenshtein distance from the input (11.90). As pointed out in Section 17.1.1.1, the framework we propose achieved the highest scores on all three TS corpora on the SAMSA metric (0.67, 0.57, 0.54), which was shown to not only highly correlate with human judgments on the syntactic correctness of the simplified sentences, but also on their structural simplicity. Therefore, we can conclude that our approach succeeded in converting the input into an output that was much simpler than the original sentences. The human evaluation conveyed further evidence that our proposed TS framework generated sentences that reached a high level of syntactic simplicity, outperforming all the baseline approaches by a large margin of at least 0.44 points. Furthermore, the frequency distribution of the transformation patterns that were triggered by our framework during the simplification process demonstrated that less than 30% of the 35 manually defined grammar rules accounted for about 75% of the rule applications, suggesting that the specified transformation patterns covered the majority of syntactic phenomena that are involved in complex sentence constructions. In combination with the descriptive statistics presented above and the human ratings from Section 15.1.2.1, which revealed that our reference TS implementation generated the shortest output sentences on average, while considerably reducing their syntactic complexity and preserving most of the information contained in the input, we deduce that our proposed structural TS approach largely succeeded in splitting complex source sentences into a sequence of atomic semantic units that presented a simplified syntax. This finding was also supported by the manual analysis of the transformation patterns that we carried out. It showed that with an average rate of more than 80% for the syntactic phenomena under consideration, the overall ratio of the simplification rules being fired whenever the corresponding clausal or phrasal construct was present in the given input sentence was very high. Finally, our manual error analysis demonstrated that only 17% of the erroneous simplifications could be attributed to additional parts that were included by mistake in a simplified proposition, i.e. they contained some optional constituents that could be discarded without leading to an incoherent or semantically meaningless output and therefore did not represent a minimal semantic unit.

Conclusion Based on the experiments we performed and the associated results, we can conclude that our proposed TS framework largely succeeds in splitting complex source sentences into a set of simplified output sentences. According to above findings, it manages to considerably reduce the syntactic complexity of the input. In fact, we were able to show that an overwhelming majority of output propositions are limited to essential components, representing minimal propositions in terms of atomic semantic units.

17.1.1.3 Hypothesis 1.3: Semantic Correctness

A complex sentence C can be transformed into a sequence of output sentences $T_1, \ldots, T_n, n >= 2$, such that the output sentences $T_1, \ldots, T_n$ convey all and only the information contained in the input.

Experimental support To examine the semantic correctness of the simplified sentences, we replicated the manual analysis for assessing the quality of the generated output based on human ratings of two judges. This time, the focus was on the question of whether the output preserves the meaning of the input. Furthermore, we calculated a variant of the SAMSA score, $SAMSA_{abl}$, which presents a high correlation with human judgments on the dimension of meaning preservation. In addition, we conducted a manual error analysis which allowed conclusions to be drawn regarding the semantic correctness of the simplified output.

Interpretation Analogous to the findings of the human evaluation presented in Section 17.1.1.1, the YATS (4.60) and RegenT (4.56) baselines slightly outperformed our proposed TS approach (4.50) in the meaning preservation ratings. However, this can be attributed again to their comparatively high reluctance in simplifying the input, which ensured that they were less prone to produce an output that obfuscated or altered the original meaning of the source sentences. The SAMSA$_{abl}$ score, which was shown to present a high correlation with human ratings on meaning preservation, conveyed further evidence that our reference TS implementation DISSIM splits the input into a simplified output that has largely the same meaning as the original sentences. In fact, our framework came no later than a close second (0.84, 0.84, 0.84) among the tested TS systems. Finally, the manual error analysis we performed identified two types of errors that frequently occurred in the transformation process, resulting in a change or disguise of the meaning of the input. These error classes included parts of a sentence that were missing in the simplified output (33%), as well as wrong referents that were assigned to an extracted span (24%).

Conclusion The results from our evaluation indicate that the rephrasings performed by our TS approach preserve most of the information contained in the input, resulting in sequences of simplified sentences that largely have the same meaning as the original sentences.

17.1.1.4 Summary

The evaluation we conducted provides sufficient corroboration evidence to support hypothesis 1.1, hypothesis 1.2 and hypothesis 1.3. Thus, we can deduce that our proposed discourse-aware TS approach succeeds in transforming sentences that present a complex linguistic structure into a sequence of simplified sentences that, to a large extent,

(i)   are grammatically well-formed,
(ii) represent minimal propositions and
(iii) preserve the meaning of the input.

17.1.2 Subtask 2: Establishing a Semantic Hierarchy

The aim of the second subtask of our discourse-aware TS approach is to set up a semantic hierarchy between the split sentences. In this context, the transformation process we propose needs to establish a contextual hierarchy between the decomposed spans *(constituency type classification)*, as well as to identify the semantic relationship that holds between each pair of split sentences *(rhetorical relation identification)*. Hence, in the following sections, we will analyze the performance of our TS framework in transforming complex sentences into hierarchically ordered and semantically interconnected simplified sentences in the light of the evaluation we conducted.

17.1.2.1 Hypothesis 1.4: Contextual Hierarchy

For each pair of structurally simplified sentences T_i and T_j, it is possible to set up a contextual hierarchy between them, distinguishing core sentences that contain the key information of the input from contextual sentences that disclose less relevant, supplementary information.

Experimental support To assess the precision of the constituency type classification step, we first mapped the simplified sentences that were generated in the sentence splitting subtask to the EDUs of the RST-DT corpus and then checked if the hierarchy of the contextual layers assigned by our TS framework corresponded to the nuclearity of the aligned text fragments of the RST-DT. In addition, we compared the performance of our TS approach with that of a set of widely used sentence-level discourse parsers on the nuclearity labeling task. To get a deeper insight into the accuracy of the contextual hierarchy established by our framework, we complemented the automatic evaluation described above by a manual analysis in which three human judges independently of each other assessed each decomposed sentence according to the following three criteria: (i) limitation to core information, (ii) soundness of the contextual proposition and (iii) correctness of the context allocation (for details, see Section 14.2.2).

Interpretation According to our experiments, the hierarchical relationship that was allocated between a pair of simplified sentences by our reference TS implementation DISSIM corresponded to the nuclearity status of the aligned EDUs from the RST-DT in 88.88% of the matched sentence pairs. Thus, in the vast majority of the cases, the hierarchy of the contextual layers assigned by our TS approach was correct. These findings were supported by our human evaluation, which revealed that 93% of the context sentences were assigned to their respective parent sentence. In addition, in more than two out of three cases, the annotators marked the propositions that were classified as core sentences by our framework as correct, thus approving that they had a meaningful interpretation and that their content was truly restricted to core information of the underlying source sentence. Finally, regarding the soundness of the context propositions, as many as 83% presented proper contextual sentences, expressing a meaningful context fact that was asserted by the input and could be properly interpreted. The comparative analysis with the discourse parser baselines, too, demonstrated the superb performance of our TS approach when it came to the task of distinguishing between nucleus and satellite spans. The evaluation results showed that our framework outperformed the baseline systems by a large margin of 13.7% at a minimum.

Conclusion Hence, the scores from both the automatic and the manual evaluation indicate that our TS framework shows a very good performance in establishing a contextual hierarchy between the split sentences by first distinguishing core sentences that contain the main information of the input from contextual sentences, whose content discloses less relevant background information, and then allocating each contextual proposition to its corresponding parent sentence.

17.1.2.2 Hypothesis 1.5: Coherence Structure

For each pair of structurally simplified sentences T_i and T_j, it is possible to preserve the coherence structure of complex sentences in the simplified output by identifying and classifying the semantic relationship that holds between two decomposed spans.

Experimental support To gauge the performance of the rhetorical relation identification step, we examined for each matching sentence pair whether the rhetorical relation assigned by our reference TS implementation DISSIM equated the relation that connected the corresponding EDUs in the RST-DT dataset. Furthermore, we determined the distribution of the relation types allocated by our TS approach when operating on the input sentences of the RST-DT and compared it to the distribution of the manually annotated rhetorical relations of this corpus. Moreover, we analyzed the performance of our TS framework on the relation labeling task in comparison to a number of discourse parser baselines. Finally, we also carried out a manual analysis in which three human annotators had to estimate whether the semantic relationship that connected a given contextual sentence to its parent was correct.

Interpretation The results from our automatic evaluation showed a decent performance of our proposed TS approach with regard to the rhetorical relation identification step. With an average precision of almost 70%, our framework succeeded in correctly identifying and classifying the rhetorical relation that held between a pair of split sentences in more than two out of three cases when run over the sentences from the RST-DT. Moreover, it again surpassed all the discourse parser baselines under consideration by at least 3.4%. Furthermore, the distribution of the rhetorical relations that were identified by our TS framework DISSIM on the source sentences of the RST-DT corpus were very similar to that of the manually annotated gold relations, suggesting that our approach performed well in accurately labeling the semantic relationships of the decomposed sentences. Finally, our human evaluation revealed that more than two-thirds of the sentence pairs were classified with the correct rhetorical relation. Only 7% of them were assigned an improper relation. However, in nearly a quarter of the cases, our TS approach was not able to identify a semantic relationship between a given pair of sentences. This can be explained by the fact that for this subtask, our framework follows a rather simplistic approach that is primarily based on cue phrases, searching for discourse markers that explicitly signal rhetorical relations. Therefore, it fails to identify a semantic relationship whenever none of the specified keywords appears in the underlying input sentence. As a result, our approach provides very precise results, but has difficulties in identifying relations that can merely be implied. This is aggravated by the fact that only a small subset of seven rhetorical relations are covered in our approach, resulting in a lack of completeness.

Conclusion Overall, the results obtained in our experiments for the rhetorical relation identification step are satisfactory. The scores indicate that our proposed TS approach is able to determine the correct semantic relationship between a given pair of sentences in more than two out of three cases on average, resulting in a coherent output where the utterances are semantically connected and, thus, present a logical structure that facilitates their correct interpretation. However, we observed some room for improvement here, in particular with respect to the semantic relationships that are not explicitly expressed in the input text in the form of cue phrases.

17.1.2.3 Summary

The evaluation provides sufficient corroboration evidence to support hypothesis 1.4 and hypothesis 1.5. Hence, we can conclude that our proposed discourse-aware TS approach not only succeeds in transforming syntactically complex sentences into a set of simplified propositions, but also to establish a semantic hierarchy between them, generating a fine-grained representation of complex assertions in the form of

(i) hierarchically ordered and
(ii) semantically interconnected sentences.

17.2 Semantically Typed Relational Tuples

Research Question 2:

Does the generated output in the form of a semantic hierarchy of minimal, self-contained propositions support machine processing in terms of Open IE applications?

17.2.1 Subtask 3: Extracting Relations and their Arguments

In an extrinsic evaluation of our discourse-aware TS framework, we investigated how the performance of state-of-the-art Open IE systems is affected when taking advantage of the semantic hierarchy of minimal propositions generated by our TS approach. For this purpose, we integrated our TS framework into existing Open IE systems as a pre-processing step for extracting semantically typed relational tuples from complex input sentences. Assuming that the fine-grained representation of complex assertions in the form of hierarchically ordered and semantically interconnected sentences with a simplified syntax improves the coverage and accuracy of the extracted tuples and supports the creation of a novel canonical context-preserving representation of relational tuples, we have examined two hypotheses, as detailed below.

17.2.1.1 Hypothesis 2.1: Coverage and Accuracy

The minimal, self-contained propositions generated by our proposed TS approach improve the performance of state-of-the-art Open IE systems in terms of precision and recall.

Experimental support In a first step, we measured if the performance of state-of-the-art Open IE approaches was improved in terms of recall and precision of the extracted relational tuples when operating on the split sentences instead of dealing directly with the raw source sentences. For this purpose, we performed a comparative analysis that examined the output of nine existing Open IE approaches on two recently proposed benchmarks. To evaluate the impact of sentence splitting on the coverage and accuracy of the relational tuples extracted by them, we compared their performance when directly operating on the raw input data with their performance when our reference TS implementation DIS-SIM was used as a pre-processing step to split complex source sentences into a set of syntactically simplified sentences with a more regular structure. To complement the automatic evaluation described above, we conducted a manual qualitative analysis of the extracted relational tuples in order to determine whether the particular deficiencies of the tested Open IE systems could be eliminated when taking advantage of the structurally simplified sentences. Aside from assessing the merits of our discourse-aware TS approach in augmenting the coverage and accuracy of the relational tuples extracted from complex input sentences in downstream Open IE applications, we evaluated the performance of our reference Open IE implementation Graphene through a comparative analysis with the same set of state-of-the-art Open IE baseline systems.

Interpretation I The results achieved in the automatic comparative evaluation conducted over a set of state-of-the-art Open IE systems demonstrated the effectiveness of using our proposed discourse-aware TS approach as a pre-processing step to improve the coverage and accuracy of the extracted relational tuples. We showed that 78% (OIE2016 benchmark) and 67% (CaRB benchmark) of the tested approaches benefit in terms of AUC and F_1 from splitting complex sentences into a set of minimal propositions with a canonical syntax, upon which the relation extraction step is then performed. Regarding the OIE2016 benchmark, the highest improvement in AUS was achieved by REVERB, yielding a +39.19% increase over the output produced when acting as a stand-alone system, while Stanford Open IE was the system with the largest gain in F_1 (+25.49%). Considering the CaRB benchmark, an increase of up to +33.33% in AUC and +24.50% in F_1, respectively, could be noted. When analyzing by hand the effects of applying our TS approach as a pre-processing step, we noted that in many cases the specific characteristics of the Open IE systems with regard to the type and boundaries of both the relational and argument phrases dissipated when operating on the simplified sentences. For instance, the overgenerated predicates that could be frequently observed in the output of OLLIE and MinIE disappeared, as well as ClausIE's overly long argument phrases.

Conclusion I Hence, we conclude that by applying our discourse-aware TS framework as a pre-processing step, Open IE systems are capable of correcting erroneous extractions, as indicated by the improved overall precision. Moreover, the higher recall scores indicate that the split sentences substantially facilitate the extraction of the relations contained in the input.

Interpretation II In addition, the analysis we performed on the OIE2016 benchmark revealed that with an average precision of 78.5%, our reference Open IE implementation Graphene succeeded in extracting highly accurate relational tuples at a high recall rate of 63.8%. Of all the Open IE systems under consideration, REVERB (83.6%) was the only approach that outperformed Graphene in terms of precision, though at the cost of a low recall score of only 44.6%. With regard to recall, RnnOIE (79.9%) was the best-performing system, setting the bar very high. Although Graphene did not achieve this target value, it was able to compete with the other high-precision Open IE systems. Regarding the other two Open IE benchmarks, CaRB and WiRe57, the overall trend was similar.

Conclusion II Thus, the results suggest that when dealing with sentences that present a simple and regular structure, a single relation extraction pattern is sufficient for extracting tuples with a high precision and recall.

17.2.1.2 Hypothesis 2.2: Canonical Context-Preserving Representation of Relational Tuples

The semantic hierarchy of minimal propositions generated by our discourse-aware TS approach

(a) *reduces the complexity of the relation extraction step, resulting in a simplistic canonical predicate-argument structure of the output, and*
(b) *allows for the enrichment of extracted relational tuples with important contextual information that supports their interpretability.*

Experimental support In a second step, we assessed the merits of our proposed discourse-aware TS approach in supporting the extraction of semantically typed relational tuples from complex assertions in downstream Open IE applications. We examined whether the semantic hierarchy of simplified sentences can be leveraged to transform the shallow semantic representation of state-of-the-art Open IE approaches into a canonical context-preserving representation of relational tuples. For this purpose, we analyzed the core characteristics of the resulting representation in the light of the experiments outlined above. The aim was to determine to what extent the semantic hierarchy of minimal propositions supports both the extraction of simplistic canonical predicate-argument structures from complex source sentences, and the enrichment of the extracted relational tuples with additional meta information to produce an output that preserves important contextual information.

Interpretation We detailed that the semantic hierarchy of minimal propositions benefits the extraction of canonical context-preserving predicate-argument structures for the task of Open IE. Representing normalized monopredicative units, the simplified sentences reduce the complexity of the relation extraction step and inherently support the extraction of canonical predicate-argument structures. Thus, a standardized output scheme is created, where each simplified sentence results in a *normalized (mostly) binary predicate-argument structure*, in which *both the predicate and the argument slots are reduced to their essential components*. In that way, the generation of overly specific predicate and argument phrases, as well as (quasi-)redundant extractions is prevented.

Moreover, we argued that the semantic hierarchy can be leveraged to incorporate important contextual information of the extracted relational tuples, thus extending the shallow semantic representation of state-of-the-art Open IE systems. We pointed out that the semantic hierarchy supports the specification of a hierarchical order between the extracted relational tuples, as it enables to distinguish between *different levels of context*—the lower the allotted layer, the more relevant is the information contained in it. In addition, we delineated that the semantic hierarchy generated by our discourse-aware TS approach can be used to enrich the output of Open IE approaches with contextual information in terms of rhetorical relations, allowing for the representation of *semantically typed relational tuples*. Thus, the extracted relations are put into a *logical structure that preserves the semantic context of the extractions*, resulting in an output that is more informative and coherent, and thus easier to interpret.

Conclusion Hence, we conclude that the semantic hierarchy of minimal propositions generated by our discourse-aware TS approach can be leveraged to transform the shallow semantic representation of existing Open IE systems into a canonical context-preserving representation of relational tuples. The proposed representation allows for a simplistic unified representation of predicate-argument structures that can easily be enriched with contextual information in terms of intra-sentential rhetorical structures and hierarchical relationships between the extracted tuples, resulting in a set of interrelated semantically typed tuples that preserve the coherence of the output.

17.2.2 Summary

The analyses outlined above provide sufficient corroboration evidence to support hypothesis 2.1 and hypothesis 2.2. Accordingly, we conclude that the semantic hierarchy of minimal propositions generated by our discourse-aware TS approach benefits downstream Open IE applications by

(i) improving their performance in terms of the precision and recall of the extracted predicate-argument structures, and

(ii) transforming the shallow semantic representation of state-of-the-art systems into a canonical context-preserving representation of relational tuples.

Limitations and Future Research Directions

The experimental results for transforming complex sentences into a semantic hierarchy of minimal propositions and leveraging this output for creating a formal semantic representation have been encouraging, showing the effectiveness of our proposed method for the three tasks of

(1) splitting complex assertions into a set of minimal propositions,
(2) establishing a semantic hierarchy between the decomposed utterances, and
(3) extracting semantically typed relational tuples.

Important short-term research challenges which became evident during the elaboration of this work are summarized below, opening directions for further research.

The results of the experiments presented in Section 15.2.1.2 showed that our TC approach was not able to identify a rhetorical relation between a pair of decomposed sentences in about a quarter of the cases. This could be attributed to the fact that the method we propose is primarily based on the detection of cue phrases, i.e. it searches the source sentences for discourse markers that explicitly signal rhetorical relations. Consequently, it has *difficulties in identifying relations that are not explicitly stated in the input, but can merely be implied.* Another limitation of the method applied in the rhetorical relation identification step is its *restriction to the determination of the seven most frequently occurring rhetorical relations* in the RST-DT. However, commonly, discourse parsers are evaluated on a much larger set of 19 classes of rhetorical relations. Thus, in the future, we plan to overcome above constraints by expanding our proposed TS approach so that implicit relationships can be recognized, too. In addition, the full set of 19 rhetorical relations that are usually employed in the area of discourse parsing shall be covered. For this purpose, we

C. Niklaus, *From Complex Sentences to a Formal Semantic Representation using Syntactic Text Simplification and Open Information Extraction*, https://doi.org/10.1007/978-3-658-38697-9_18

aim to develop a data-driven approach, following recent discourse parsing methods (e.g., Feng and Hirst (2014), Joty et al. (2015), Liu and Lapata (2017), Lin et al. (2019)) that use supervised approaches which are based on a rich set of features to complement the lookup of cue phrases.

Moreover, as of now, our TS framework operates on the level of individual sentences. Accordingly, it only considers intra-sentential relationships when creating the semantic hierarchy of minimal propositions. A promising future research direction will be to examine how the approach we propose can be extended to capture *inter-sentence relationships*, resulting in a linked proposition tree that covers not only a single source sentence, but rather a paragraph or even a full document, similar to rhetorical structure trees generated by RST parsers. Including inter-sentential semantic relations would allow to consider the wider semantic context of a proposition, beyond a single sentence. In that way, a simplified utterance could be embedded in the context of a whole text, making it possible to draw broader conclusions.

In addition, we aim to extend our proposed discourse-aware TS approach with respect to

(i) *other domains*: Currently, the application of a lexicalized parser throughout the process of splitting complex input sentences into a simplified output ties our TS framework to general-purpose texts, such as newswire and Wikipedia articles. To expand our approach to cover further important textual areas that frequently deal with complex assertions, including for example biomedical and legal texts, additional research needs to be carried out.

(ii) *other downstream applications*: As of now, we focus on the task of Open IE as an example of a downstream NLP application that benefits from the fine-grained representation of complex assertions in the form of a semantic hierarchy of minimal propositions. In the future, we plan to examine to what extent other NLP tasks, such as MT or Text Summarization, may take advantage of the output generated by our TS approach.

(iii) *other languages*: To date, the TS framework we propose is limited to English texts. Prospectively, we intend to port the idea of creating a hierarchical representation of semantically-linked minimal propositions to other languages. However, due to differences in the syntax, parse tree representations and language-specific phenomena, the specified rules cannot be directly mapped to a new language. Instead, a careful analysis of the transformation patterns and their portability to other languages is required to figure out which adaptions are necessary to transfer them to languages other than English.

Finally, enriching the extracted relational tuples with further meta information is another promising research avenue. For this purpose, we aim to *expand the lightweight semantic representation* for Open IE that we proposed in this thesis. As of now, it is restricted to capturing inter-proposition rhetorical structures only. However, this does not allow for a more detailed semantic analysis at the level of individual propositions to further characterize a single event that is expressed by a relational tuple. By assigning thematic roles in the style of SRL to the tuples' arguments, the informativeness of the extracted relations may be increased. In that way, we can indicate exactly what semantic relations hold among a predicate and its associated arguments. Accordingly, the participants in an event, as well as other event properties can be further semantically classified, helping to better understand the exact meaning conveyed in the extracted relational tuples. Thus, we improve the use of the semantic representation generated by our framework for downstream NLP applications.

Bibliography

Omri Abend and Ari Rappoport. 2013. Universal conceptual cognitive annotation (UCCA). In *Proceedings of the 51st Annual Meeting of the Association for Computational Linguistics (Volume 1: Long Papers)*, pages 228–238, Sofia, Bulgaria. Association for Computational Linguistics. https://www.aclweb.org/anthology/P13-1023

Omri Abend and Ari Rappoport. 2017. The state of the art in semantic representation. In *Proceedings of the 55th Annual Meeting of the Association for Computational Linguistics (Volume 1: Long Papers)*, pages 77–89, Vancouver, Canada. Association for Computational Linguistics. https://doi.org/10.18653/v1/P17-1008

Roberta G Abraham. 1985. Field independence-dependence and the teaching of grammar. *Tesol Quarterly*, 19(4):689–702.

Eugene Agichtein and Luis Gravano. 2000. Snowball: Extracting relations from large plain-text collections. In *Proceedings of the Fifth ACM Conference on Digital Libraries*, DL '00, page 85–94, New York, NY, USA. Association for Computing Machinery. https://doi.org/10.1145/336597.336644

Roee Aharoni and Yoav Goldberg. 2018. Split and rephrase: Better evaluation and stronger baselines. In *Proceedings of the 56th Annual Meeting of the Association for Computational Linguistics (Volume 2: Short Papers)*, pages 719–724, Melbourne, Australia. Association for Computational Linguistics. https://doi.org/10.18653/v1/P18-2114

Alan Akbik and Alexander Löser. 2012. KrakeN: N-ary facts in open information extraction. In *Proceedings of the Joint Workshop on Automatic Knowledge Base Construction and Web-scale Knowledge Extraction (AKBC-WEKEX)*, pages 52–56, Montréal, Canada. Association for Computational Linguistics. https://www.aclweb.org/anthology/W12-3010

Fernando Alva-Manchego, Carolina Scarton, and Lucia Specia. 2020. Data-driven sentence simplification: Survey and benchmark. *Computational Linguistics*, 46(1):135–187. https://doi.org/10.1162/coli_a_00370

Fernando Alva-Manchego, Joachim Bingel, Gustavo Paetzold, Carolina Scarton, and Lucia Specia. 2017. Learning how to simplify from explicit labeling of complex-simplified text pairs. In *Proceedings of the Eighth International Joint Conference on Natural Language Processing (Volume 1: Long Papers)*, pages 295–305, Taipei, Taiwan. Asian Federation of Natural Language Processing. https://www.aclweb.org/anthology/I17-1030

Fernando Alva-Manchego, Louis Martin, Antoine Bordes, Carolina Scarton, Benoît Sagot, and Lucia Specia. 2020. ASSET: A dataset for tuning and evaluation of sentence simplification models with multiple rewriting transformations. In *Proceedings of the 58th Annual*

© The Editor(s) (if applicable) and The Author(s), under exclusive license to Springer Fachmedien Wiesbaden GmbH, part of Springer Nature 2022
C. Niklaus, *From Complex Sentences to a Formal Semantic Representation using Syntactic Text Simplification and Open Information Extraction*,
https://doi.org/10.1007/978-3-658-38697-9

Meeting of the Association for Computational Linguistics, pages 4668–4679, Online. Association for Computational Linguistics. https://www.aclweb.org/anthology/2020.acl-main. 424

Gabor Angeli, Melvin Jose Johnson Premkumar, and Christopher D. Manning. 2015. Leveraging linguistic structure for open domain information extraction. In *Proceedings of the 53rd Annual Meeting of the Association for Computational Linguistics and the 7th International Joint Conference on Natural Language Processing (Volume 1: Long Papers)*, pages 344–354, Beijing, China. Association for Computational Linguistics. https://doi.org/10.3115/v1/P15-1034

Mandya Angrosh, Tadashi Nomoto, and Advaith Siddharthan. 2014. Lexico-syntactic text simplification and compression with typed dependencies. In *Proceedings of COLING 2014, the 25th International Conference on Computational Linguistics: Technical Papers*, pages 1996–2006, Dublin, Ireland. Dublin City University and Association for Computational Linguistics. https://www.aclweb.org/anthology/C14-1188

Nguyen Bach, Qin Gao, Stephan Vogel, and Alex Waibel. 2011. TriS: A statistical sentence simplifier with log-linear models and margin-based discriminative training. In *Proceedings of 5th International Joint Conference on Natural Language Processing*, pages 474–482, Chiang Mai, Thailand. Asian Federation of Natural Language Processing. https://www.aclweb.org/anthology/I11-1053

Dzmitry Bahdanau, Kyunghyun Cho, and Yoshua Bengio. 2015. Neural machine translation by jointly learning to align and translate. In *3rd International Conference on Learning Representations, ICLR 2015, San Diego, CA, USA, May 7–9, 2015, Conference Track Proceedings*. http://arxiv.org/abs/1409.0473

Daniel Bailey, Yuliya Lierler, and Benjamin Susman. 2015. Prepositional phrase attachment problem revisited: how verbnet can help. In *Proceedings of the 11th International Conference on Computational Semantics*, pages 12–22, London, UK. Association for Computational Linguistics. https://www.aclweb.org/anthology/W15-0102

Laura Banarescu, Claire Bonial, Shu Cai, Madalina Georgescu, Kira Griffitt, Ulf Hermjakob, Kevin Knight, Philipp Koehn, Martha Palmer, and Nathan Schneider. 2013. Abstract Meaning Representation for sembanking. In *Proceedings of the 7th Linguistic Annotation Workshop and Interoperability with Discourse*, pages 178–186, Sofia, Bulgaria. Association for Computational Linguistics. https://www.aclweb.org/anthology/W13-2322

Michele Banko, Michael J. Cafarella, Stephen Soderland, Matt Broadhead, and Oren Etzioni. 2007. Open information extraction from the web. In *Proceedings of the 20th International Joint Conference on Artifical Intelligence*, pages 2670–2676, San Francisco, CA, USA. Morgan Kaufmann Publishers Inc.

Regina Barzilay and Noemie Elhadad. 2003. Sentence alignment for monolingual comparable corpora.In *Proceedings of the 2003 Conference on Empirical Methods in Natural Language Processing*, pages 25–32. https://www.aclweb.org/anthology/W03-1004

Hannah Bast and Elmar Haussmann. 2013. Open information extraction via contextual sentence decomposition. In *2013 IEEE Seventh International Conference on Semantic Computing*, pages 154–159. IEEE.

Beata Beigman Klebanov, Kevin Knight, and Daniel Marcu. 2004. Text simplification for information-seeking applications. In *On the Move to Meaningful Internet Systems 2004: CoopIS, DOA, and ODBASE*, pages 735–747, Berlin, Heidelberg. Springer Berlin Heidelberg.

Farah Benamara and Maite Taboada. 2015. Mapping different rhetorical relation annotations: A proposal. In *Proceedings of the Fourth Joint Conference on Lexical and Computational Semantics*, pages 147–152, Denver, Colorado. Association for Computational Linguistics. https://doi.org/10.18653/v1/S15-1016

Delphine Bernhard, Louis De Viron, Véronique Moriceau, and Xavier Tannier. 2012. Question generation for french: collating parsers and paraphrasing questions. *Dialogue & Discourse*, 3(2):43–74.

Sangnie Bhardwaj, Samarth Aggarwal, and Mausam Mausam. 2019. CaRB: A crowdsourced benchmark for open IE. In *Proceedings of the 2019 Conference on Empirical Methods in Natural Language Processing and the 9th International Joint Conference on Natural Language Processing (EMNLP-IJCNLP)*, pages 6262–6267, Hong Kong, China. Association for Computational Linguistics. https://doi.org/10.18653/v1/D19-1651

Nikita Bhutani, H. V. Jagadish, and Dragomir Radev. 2016. Nested propositions in open information extraction. In *Proceedings of the 2016 Conference on Empirical Methods in Natural Language Processing*, pages 55–64, Austin, Texas. Association for Computational Linguistics. https://doi.org/10.18653/v1/D16-1006

Or Biran, Samuel Brody, and Noémie Elhadad. 2011. Putting it simply: a context-aware approach to lexical simplification. In *Proceedings of the 49th Annual Meeting of the Association for Computational Linguistics: Human Language Technologies*, pages 496–501, Portland, Oregon, USA. Association for Computational Linguistics. https://www.aclweb.org/anthology/P11-2087

Jan A. Botha, Manaal Faruqui, John Alex, Jason Baldridge, and Dipanjan Das. 2018. Learning to split and rephrase from Wikipedia edit history. In *Proceedings of the 2018 Conference on Empirical Methods in Natural Language Processing*, pages 732–737, Brussels, Belgium. Association for Computational Linguistics. https://doi.org/10.18653/v1/D18-1080

Nadjet Bouayad-Agha, Gerard Casamayor, Gabriela Ferraro, Simon Mille, Vanesa Vidal, and Leo Wanner. 2009. Improving the comprehension of legal documentation: The case of patent claims. In *Proceedings of the 12th International Conference on Artificial Intelligence and Law*, ICAIL '09, page 78–87, New York, NY, USA. Association for Computing Machinery. https://doi.org/10.1145/1568234.1568244

Sergey Brin. 1998. Extracting patterns and relations from the world wide web. In *Selected Papers from the International Workshop on The World Wide Web and Databases*, WebDB '98, page 172–183, Berlin, Heidelberg. Springer-Verlag.

Laurel J. Brinton. 2000. *The Structure of Modern English: A linguistic introduction*. John Benjamins B.V., Amsterdam, The Netherlands.

Laetitia Brouwers, Delphine Bernhard, Anne-Laure Ligozat, and Thomas François. 2014. Syntactic sentence simplification for French. In *Proceedings of the 3rd Workshop on Predicting and Improving Text Readability for Target Reader Populations (PITR)*, pages 47–56, Gothenburg, Sweden. Association for Computational Linguistics. https://doi.org/10.3115/v1/W14-1206

Lynn Carlson and Daniel Marcu. 2001. Discourse tagging reference manual. *ISI Technical Report ISI-TR-545*, 54:56. https://www.isi.edu/~marcu/discourse/tagging-ref-manual.pdf

Lynn Carlson, Mary Ellen Okurowski, and Daniel Marcu. 2002. *RST discourse treebank*. Linguistic Data Consortium, University of Pennsylvania.

John Carroll, Guido Minnen, Darren Pearce, Yvonne Canning, Siobhan Devlin, and John Tait. 1999. Simplifying text for language-impaired readers. In *Ninth Conference of the European Chapter of the Association for Computational Linguistics*.

Matthias Cetto. 2017. *Rhetorical, Clausal and Phrasal Disembedding for Open Relation Extraction*. Master thesis, University of Passau.

Matthias Cetto, Christina Niklaus, André Freitas, and Siegfried Handschuh. 2018a. Graphene: a context-preserving open information extraction system. In *Proceedings of the 27th International Conference on Computational Linguistics: System Demonstrations*, pages 94–98, Santa Fe, New Mexico. Association for Computational Linguistics. https://www.aclweb.org/anthology/C18-2021

Matthias Cetto, Christina Niklaus, André Freitas, and Siegfried Handschuh. 2018b. Graphene: Semantically-linked propositions in open information extraction. In *Proceedings of the 27th International Conference on Computational Linguistics*, pages 2300–2311, Santa Fe, New Mexico, USA. Association for Computational Linguistics. https://www.aclweb.org/anthology/C18-1195

R. Chandrasekar, Christine Doran, and B. Srinivas. 1996. Motivations and methods for text simplification. In *COLING 1996 Volume 2: The 16th International Conference on Computational Linguistics*. https://www.aclweb.org/anthology/C96-2183

Kyunghyun Cho, Bart van Merriënboer, Caglar Gulcehre, Dzmitry Bahdanau, Fethi Bougares, Holger Schwenk, and Yoshua Bengio. 2014. Learning phrase representations using RNN encoder–decoder for statistical machine translation. In *Proceedings of the 2014 Conference on Empirical Methods in Natural Language Processing (EMNLP)*, pages 1724–1734, Doha, Qatar. Association for Computational Linguistics. https://doi.org/10.3115/v1/D14-1179

Janara Christensen, Mausam, Stephen Soderland, and Oren Etzioni. 2010. Semantic role labeling for open information extraction. In *Proceedings of the NAACL HLT 2010 First International Workshop on Formalisms and Methodology for Learning by Reading*, pages 52–60, Los Angeles, California. Association for Computational Linguistics. https://www.aclweb.org/anthology/W10-0907

Jacob Cohen. 1968. Weighted kappa: Nominal scale agreement provision for scaled disagreement or partial credit. *Psychological bulletin*, 70(4):213.

Trevor Cohn and Mirella Lapata. 2009. Sentence compression as tree transduction. *Journal of Artificial Intelligence Research*, 34(1):637–674.

Will Coster and David Kauchak. 2011a. Learning to simplify sentences using Wikipedia. In *Proceedings of the Workshop on Monolingual Text-To-Text Generation*, pages 1–9, Portland, Oregon. Association for Computational Linguistics. https://www.aclweb.org/anthology/W11-1601

William Coster and David Kauchak. 2011b. Simple English Wikipedia: A new text simplification task. In *Proceedings of the 49th Annual Meeting of the Association for Computational Linguistics: Human Language Technologies*, pages 665–669, Portland, Oregon, USA. Association for Computational Linguistics. https://www.aclweb.org/anthology/P11-2117

Lei Cui, Furu Wei, and Ming Zhou. 2018. Neural open information extraction. In *Proceedings of the 56th Annual Meeting of the Association for Computational Linguistics (Volume 2: Short Papers)*, pages 407–413, Melbourne, Australia. Association for Computational Linguistics. https://doi.org/10.18653/v1/P18-2065

James Curran, Stephen Clark, and Johan Bos. 2007. Linguistically motivated large-scale NLP with c&c and boxer. In *Proceedings of the 45th Annual Meeting of the Association for Computational Linguistics Companion Volume Proceedings of the Demo and Poster Sessions*, pages 33–36, Prague, Czech Republic. Association for Computational Linguistics. https://www.aclweb.org/anthology/P07-2009

Jan De Belder and Marie-Francine Moens. 2010. Text simplification for children. In *Proceedings of the SIGIR workshop on accessible search systems*, pages 19–26. ACM; New York. https://lirias.kuleuven.be/retrieve/120012 Main article [freely available]

Luciano Del Corro and Rainer Gemulla. 2013. Clausie: Clause-based open information extraction. In *Proceedings of the 22nd International Conference on World Wide Web*, WWW '13, page 355–366, New York, NY, USA. Association for Computing Machinery. https://doi.org/10.1145/2488388.2488420

Jacob Devlin, Ming-Wei Chang, Kenton Lee, and Kristina Toutanova. 2019. BERT: Pretraining of deep bidirectional transformers for language understanding. In *Proceedings of the 2019 Conference of the North American Chapter of the Association for Computational Linguistics: Human Language Technologies, Volume 1 (Long and Short Papers)*, pages 4171–4186, Minneapolis, Minnesota. Association for Computational Linguistics. https://doi.org/10.18653/v1/N19-1423

Robert MW Dixon. 2010a. *Basic Linguistic Theory Volume 1: Methodology*, volume 1.Oxford University Press.

Robert MW Dixon. 2010b. *Basic Linguistic Theory Volume 2: Grammatical Topics*, volume 2. Oxford University Press.

Robert MW Dixon. 2012. *Basic Linguistic Theory Volume 3: Further Grammatical Topics*, volume 3. Oxford University Press.

Yue Dong, Zichao Li, Mehdi Rezagholizadeh, and Jackie Chi Kit Cheung. 2019. EditNTS: An neural programmer-interpreter model for sentence simplification through explicit editing. In *Proceedings of the 57th Annual Meeting of the Association for Computational Linguistics*, pages 3393–3402, Florence, Italy. Association for Computational Linguistics. https://doi.org/10.18653/v1/P19-1331

David Dowty. 1991. Thematic proto-roles and argument selection. *Language*, 67(3):547–619. http://www.jstor.org/stable/415037

Jan van Eijck. 1990. *Discourse representation theory*. Citeseer.

Jason Eisner. 2003. Learning non-isomorphic tree mappings for machine translation. In *The Companion Volume to the Proceedings of 41st Annual Meeting of the Association for Computational Linguistics*, pages 205–208, Sapporo, Japan. Association for Computational Linguistics. https://doi.org/10.3115/1075178.1075217

Richard Evans, Constantin Orăsan, and Iustin Dornescu. 2014. An evaluation of syntactic simplification rules for people with autism. In *Proceedings of the 3rd Workshop on Predicting and Improving Text Readability for Target Reader Populations (PITR)*, pages 131–140, Gothenburg, Sweden. Association for Computational Linguistics. https://doi.org/10.3115/v1/W14-1215

Richard Evans and Constantin Orăsan. 2019. Identifying signs of syntactic complexity for rule-based sentence simplification. *Natural Language Engineering*, 25(1):69–119. https://doi.org/10.1017/S1351324918000384

Richard J Evans. 2011. Comparing methods for the syntactic simplification of sentences in information extraction. *Literary and linguistic computing*, 26(4):371–388.

Anthony Fader, Stephen Soderland, and Oren Etzioni. 2011. Identifying relations for open information extraction. In *Proceedings of the 2011 Conference on Empirical Methods in Natural Language Processing*, pages 1535–1545, Edinburgh, Scotland, UK. Association for Computational Linguistics. https://www.aclweb.org/anthology/D11-1142

Richard Fay, editor. 1990. *Collins Cobuild English Grammar*. Collins.

Dan Feblowitz and David Kauchak. 2013. Sentence simplification as tree transduction. In *Proceedings of the Second Workshop on Predicting and Improving Text Readability for Target Reader Populations*, pages 1–10, Sofia, Bulgaria. Association for Computational Linguistics. https://www.aclweb.org/anthology/W13-2901

Vanessa Wei Feng and Graeme Hirst. 2014. A linear-time bottom-up discourse parser with constraints and post-editing. In *Proceedings of the 52nd Annual Meeting of the Association for Computational Linguistics (Volume 1: Long Papers)*, pages 511–521, Baltimore, Maryland. Association for Computational Linguistics. https://doi.org/10.3115/v1/P14-1048

Daniel Ferrés, Montserrat Marimon, Horacio Saggion, and Ahmed AbuRa'ed. 2016. YATS: yet another text simplifier. In *Natural Language Processing and Information Systems—21st International Conference on Applications of Natural Language to Information Systems, NLDB 2016, Salford, UK, June 22–24, 2016, Proceedings*, volume 9612 of *Lecture Notes in Computer Science*, pages 335–342. Springer. https://doi.org/10.1007/978-3-319-41754-7_32

Katja Filippova and Michael Strube. 2008. Sentence fusion via dependency graph compression. In *Proceedings of the 2008 Conference on Empirical Methods in Natural Language Processing*, pages 177–185, Honolulu, Hawaii. Association for Computational Linguistics. https://www.aclweb.org/anthology/D08-1019

Charles J. Fillmore, Josef Ruppenhofer, and Collin F. Baker. 2004. *FrameNet and Representing the Link between Semantic and Syntactic Relations*, Language and Linguistics Monographs Series B, pages 19–62. Institute of Linguistics, Academia Sinica, Taipei.

Seeger Fisher and Brian Roark. 2007. The utility of parse-derived features for automatic discourse segmentation. In *Proceedings of the 45th Annual Meeting of the Association of Computational Linguistics*, pages 488–495, Prague, Czech Republic. Association for Computational Linguistics. https://www.aclweb.org/anthology/P07-1062

J.L. Fleiss et al. 1971. Measuring nominal scale agreement among many raters. *Psychological Bulletin*, 76(5):378–382.

Juri Ganitkevitch, Benjamin Van Durme, and Chris Callison-Burch. 2013. PPDB: The paraphrase database. In *Proceedings of the 2013 Conference of the North American Chapter of the Association for Computational Linguistics: Human Language Technologies*, pages 758–764, Atlanta, Georgia. Association for Computational Linguistics. https://www.aclweb.org/anthology/N13-1092

Claire Gardent, Anastasia Shimorina, Shashi Narayan, and Laura Perez-Beltrachini. 2017. The WebNLG challenge: Generating text from RDF data. In *Proceedings of the 10th International Conference on Natural Language Generation*, pages 124–133, Santiago de Compostela, Spain. Association for Computational Linguistics. https://doi.org/10.18653/v1/W17-3518

Kiril Gashteovski, Rainer Gemulla, and Luciano del Corro. 2017. MinIE: Minimizing facts in open information extraction. In *Proceedings of the 2017 Conference on Empirical Methods in Natural Language Processing*, pages 2630–2640, Copenhagen, Denmark. Association for Computational Linguistics. https://doi.org/10.18653/v1/D17-1278

Kiril Gashteovski, Sebastian Wanner, Sven Hertling, Samuel Broscheit, and Rainer Gemulla. 2019. OPIEC: an open information extraction corpus. In *1st Conference on Automated Knowledge Base Construction, AKBC 2019, Amherst, MA, USA, May 20–22, 2019*. https://doi.org/10.24432/C53W2J

Alexander Gelbukh and Hiram Calvo. 2018. *Prepositional Phrase Attachment Disambiguation*, pages 85–110. Springer International Publishing, Cham.

Goran Glavaš and Sanja Štajner. 2015. Simplifying lexical simplification: Do we need simplified corpora? In *Proceedings of the 53rd Annual Meeting of the Association for Computational Linguistics and the 7th International Joint Conference on Natural Language Processing (Volume 2: Short Papers)*, pages 63–68, Beijing, China. Association for Computational Linguistics. https://doi.org/10.3115/v1/P15-2011

Jiatao Gu, Zhengdong Lu, Hang Li, and Victor O.K. Li. 2016. Incorporating copying mechanism in sequence-to-sequence learning. In *Proceedings of the 54th Annual Meeting of the Association for Computational Linguistics (Volume 1: Long Papers)*, pages 1631–1640, Berlin, Germany. Association for Computational Linguistics. https://doi.org/10.18653/v1/P16-1154

Han Guo, Ramakanth Pasunuru, and Mohit Bansal. 2018. Dynamic multi-level multi-task learning for sentence simplification. In *Proceedings of the 27th International Conference on Computational Linguistics*, pages 462–476, Santa Fe, New Mexico, USA. Association for Computational Linguistics. https://www.aclweb.org/anthology/C18-1039

Luheng He, Kenton Lee, Mike Lewis, and Luke Zettlemoyer. 2017. Deep semantic role labeling: What works and what's next. In *Proceedings of the 55th Annual Meeting of the Association for Computational Linguistics (Volume 1: Long Papers)*, pages 473–483, Vancouver, Canada. Association for Computational Linguistics. https://doi.org/10.18653/v1/P17-1044

Luheng He, Mike Lewis, and Luke Zettlemoyer. 2015. Question-answer driven semantic role labeling: Using natural language to annotate natural language. In *Proceedings of the 2015 Conference on Empirical Methods in Natural Language Processing*, pages 643–653, Lisbon, Portugal. Association for Computational Linguistics. https://doi.org/10.18653/v1/D15-1076

Michael Heilman and Noah A. Smith. 2010. Extracting simplified statements for factual question generation. In *Proceedings of the QG2010: The Third Workshop on Question Generation*, pages 11–20.

Hugo Hernault, Helmut Prendinger, Mitsuru Ishizuka, et al. 2010. Hilda: A discourse parser using support vector machine classification. *Dialogue & Discourse*, 1(3).

Daniel Hershcovich, Omri Abend, and Ari Rappoport. 2017. A transition-based directed acyclic graph parser for UCCA. In *Proceedings of the 55th Annual Meeting of the Association for Computational Linguistics (Volume 1: Long Papers)*, pages 1127–1138, Vancouver, Canada. Association for Computational Linguistics. https://doi.org/10.18653/v1/P17-1104

Sepp Hochreiter and Jürgen Schmidhuber. 1997. Long short-term memory. *Neural Computation*, 9(8):1735–1780. https://doi.org/10.1162/neco.1997.9.8.1735

William Hwang, Hannaneh Hajishirzi, Mari Ostendorf, and Wei Wu. 2015. Aligning sentences from standard Wikipedia to simple Wikipedia. In *Proceedings of the 2015 Conference of the North American Chapter of the Association for Computational Linguistics: Human Language Technologies*, pages 211–217, Denver, Colorado. Association for Computational Linguistics. https://doi.org/10.3115/v1/N15-1022

Kentaro Inui, Atsushi Fujita, Tetsuro Takahashi, Ryu Iida, and Tomoya Iwakura. 2003. Text simplification for reading assistance: A project note. In *Proceedings of the Second International Workshop on Paraphrasing—Volume 16*, PARAPHRASE '03, page 9–16, USA. Association for Computational Linguistics. https://doi.org/10.3115/1118984.1118986

Yangfeng Ji and Jacob Eisenstein. 2014. Representation learning for text-level discourse parsing. In *Proceedings of the 52nd Annual Meeting of the Association for Computational Linguistics (Volume 1: Long Papers)*, pages 13–24, Baltimore, Maryland. Association for Computational Linguistics. https://doi.org/10.3115/v1/P14-1002

Chao Jiang, Mounica Maddela, Wuwei Lan, Yang Zhong, and Wei Xu. 2020. Neural CRF model for sentence alignment in text simplification. In *Proceedings of the 58th Annual Meeting of the Association for Computational Linguistics*, pages 7943–7960, Online. Association for Computational Linguistics. https://doi.org/10.18653/v1/2020.acl-main.709

Shafiq Joty, Giuseppe Carenini, and Raymond T. Ng. 2015. CODRA: A novel discriminative framework for rhetorical analysis. *Computational Linguistics*, 41(3):385–435. https://doi.org/10.1162/COLI_a_00226

Daniel Jurafsky and James H. Martin. 2009. *Speech and Language Processing (2Nd Edition)*. Prentice-Hall Inc, Upper Saddle River, NJ, USA.

Tomoyuki Kajiwara and Mamoru Komachi. 2016. Building a monolingual parallel corpus for text simplification using sentence similarity based on alignment between word embeddings. In *Proceedings of COLING 2016, the 26th International Conference on Computational Linguistics: Technical Papers*, pages 1147–1158, Osaka, Japan. The COLING 2016 Organizing Committee. https://www.aclweb.org/anthology/C16-1109

Hans Kamp. 1981. A theory of truth and semantic representation. *Formal semantics-the essential readings*, pages 189–222.

Hans Kamp and Uwe Reyle. 1993. *From Discourse to Logic: Introduction to modeltheoretic semantics of natural language, formal logic and Discourse Representation Theory*. Kluwer Academic Publishers.

Hans Kamp and Christian Rohrer. 1983. Tense in texts. In Rainer Bäuerle, Christoph Schwarze, and Arnim von Stechow, editors, *Meaning, Use, and Interpretation of Language*, pages 250–269. de Gruyter. https://doi.org/10.1515/9783110852820.250

David Kauchak. 2013. Improving text simplification language modeling using unsimplified text data. In *Proceedings of the 51st Annual Meeting of the Association for Computational Linguistics (Volume 1: Long Papers)*, pages 1537–1546, Sofia, Bulgaria. Association for Computational Linguistics. https://www.aclweb.org/anthology/P13-1151

Kate Kearns. 2011. *Semantics*. Palgrave Modern Linguistics. Palgrave Macmillan. https://books.google.de/books?id=kK0cBQAAQBAJ

Andrej A Kibrik and Olga N Krasavina. 2005. A corpus study of referential choice: The role of rhetorical structure. *Proceedings of DIALOG'05*.

J Peter Kincaid, Robert P Fishburne Jr, Richard L Rogers, and Brad S Chissom. 1975. Derivation of new readability formulas (automated readability index, fog count and flesch reading ease formula) for navy enlisted personnel. Technical report, Naval Technical Training Command Millington TN Research Branch.

Karin Kipper, Hoa Trang Dang, and Martha Palmer. 2000. Class-based construction of a verb lexicon. In *Proceedings of the Seventeenth National Conference on Artificial Intelligence and Twelfth Conference on Innovative Applications of Artificial Intelligence*, page 691–696. AAAI Press.

Guillaume Klein, Yoon Kim, Yuntian Deng, Jean Senellart, and Alexander Rush. 2017. Open-NMT: Open-source toolkit for neural machine translation. In *Proceedings of ACL 2017, System Demonstrations*, pages 67–72, Vancouver, Canada. Association for Computational Linguistics. https://www.aclweb.org/anthology/P17-4012

Alistair Knott and Robert Dale. 1994. Using linguistic phenomena to motivate a set of coherence relations. *Discourse processes*, 18(1):35–62.

Philipp Koehn, Hieu Hoang, Alexandra Birch, Chris Callison-Burch, Marcello Federico, Nicola Bertoldi, Brooke Cowan, Wade Shen, Christine Moran, Richard Zens, Chris Dyer, Ondřej Bojar, Alexandra Constantin, and Evan Herbst. 2007. Moses: Open source toolkit for statistical machine translation. In *Proceedings of the 45th Annual Meeting of the Association for Computational Linguistics Companion Volume Proceedings of the Demo and Poster Sessions*, pages 177–180, Prague, Czech Republic. Association for Computational Linguistics. https://www.aclweb.org/anthology/P07-2045

Reno Kriz, João Sedoc, Marianna Apidianaki, Carolina Zheng, Gaurav Kumar, Eleni Miltsakaki, and Chris Callison-Burch. 2019. Complexity-weighted loss and diverse reranking for sentence simplification. In *Proceedings of the 2019 Conference of the North American Chapter of the Association for Computational Linguistics: Human Language Technologies, Volume 1 (Long and Short Papers)*, pages 3137–3147, Minneapolis, Minnesota. Association for Computational Linguistics. https://doi.org/10.18653/v1/N19-1317

Matt J. Kusner, Yu Sun, Nicholas I. Kolkin, and Kilian Q. Weinberger. 2015. From word embeddings to document distances. In *Proceedings of the 32nd International Conference on International Conference on Machine Learning—Volume 37*, ICML'15, page 957–966. JMLR.org.

John D. Lafferty, Andrew McCallum, and Fernando C. N. Pereira. 2001. Conditional random fields: Probabilistic models for segmenting and labeling sequence data. In *Proceedings of the Eighteenth International Conference on Machine Learning*, ICML '01, page 282–289, San Francisco, CA, USA. Morgan Kaufmann Publishers Inc.

Ronald W Langacker. 2008. *Cognitive Grammar: A Basic Introduction*. Oxford University Press, USA.

Alex Lascarides and Nicholas Asher. 2007. *Segmented Discourse Representation Theory: Dynamic Semantics With Discourse Structure*, pages 87–124. Springer Netherlands, Dordrecht. https://doi.org/10.1007/978-1-4020-5958-2_5

William Lechelle, Fabrizio Gotti, and Phillippe Langlais. 2019. WiRe57: A fine-grained benchmark for open information extraction. In *Proceedings of the 13th Linguistic Annotation Workshop*, pages 6–15, Florence, Italy. Association for Computational Linguistics. https://doi.org/10.18653/v1/W19-4002

Roger Levy and Galen Andrew. 2006. Tregex and tsurgeon: tools for querying and manipulating tree data structures. In *Proceedings of the Fifth International Conference on Language Resources and Evaluation (LREC'06)*, Genoa, Italy. European Language Resources Association (ELRA). http://www.lrec-conf.org/proceedings/lrec2006/pdf/513_pdf.pdf

Jiwei Li, Rumeng Li, and Eduard Hovy. 2014. Recursive deep models for discourse parsing. In *Proceedings of the 2014 Conference on Empirical Methods in Natural Language Processing (EMNLP)*, pages 2061–2069, Doha, Qatar. Association for Computational Linguistics. https://doi.org/10.3115/v1/D14-1220

Xiang Lin, Shafiq Joty, Prathyusha Jwalapuram, and M Saiful Bari. 2019. A unified linear-time framework for sentence-level discourse parsing. In *Proceedings of the 57th Annual*

Meeting of the Association for Computational Linguistics, pages 4190–4200, Florence, Italy. Association for Computational Linguistics. https://doi.org/10.18653/v1/P19-1410

Jiangming Liu, Shay B. Cohen, and Mirella Lapata. 2019. Discourse representation parsing for sentences and documents. In *Proceedings of the 57th Annual Meeting of the Association for Computational Linguistics*, pages 6248–6262, Florence, Italy. Association for Computational Linguistics. https://doi.org/10.18653/v1/P19-1629

Yang Liu and Mirella Lapata. 2017. Learning contextually informed representations for linear-time discourse parsing. In *Proceedings of the 2017 Conference on Empirical Methods in Natural Language Processing*, pages 1289–1298, Copenhagen, Denmark. Association for Computational Linguistics. https://doi.org/10.18653/v1/D17-1133

Jonathan Mallinson and Mirella Lapata. 2019. Controllable sentence simplification: Employing syntactic and lexical constraints. *arXiv preprint* arXiv:1910.04387.

William C Mann and Sandra A Thompson. 1988. Rhetorical structure theory: Toward a functional theory of text organization. *Text-Interdisciplinary Journal for the Study of Discourse*, 8(3), 243–281.

Daniel Marcu. 1997. The rhetorical parsing of unrestricted natural language texts. In *35th Annual Meeting of the Association for Computational Linguistics and 8th Conference of the European Chapter of the Association for Computational Linguistics*, pages 96–103, Madrid, Spain. Association for Computational Linguistics. https://doi.org/10.3115/976909.979630

Daniel Marcu. 2000. The rhetorical parsing of unrestricted texts: a surface-based approach. *Computational Linguistics*, 26(3):395–448. https://www.aclweb.org/anthology/J00-3005

Daniel C. Marcu. 1998. *The Rhetorical Parsing, Summarization, and Generation of Natural Language Texts*. Ph.D. thesis, University of Toronto, CAN. AAINQ35238.

Mitchell P. Marcus, Beatrice Santorini, and Mary Ann Marcinkiewicz. 1993. Building a large annotated corpus of English: The Penn Treebank. *Computational Linguistics*, 19(2):313–330. https://www.aclweb.org/anthology/J93-2004

Marie-Catherine de Marneffe, Timothy Dozat, Natalia Silveira, Katri Haverinen, Filip Ginter, Joakim Nivre, and Christopher D. Manning. 2014. Universal Stanford dependencies: A cross-linguistic typology. In *Proceedings of the Ninth International Conference on Language Resources and Evaluation (LREC'14)*, pages 4585–4592, Reykjavik, Iceland. European Language Resources Association (ELRA). http://www.lrec-conf.org/proceedings/lrec2014/pdf/1062_Paper.pdf

Lluís Màrquez, Xavier Carreras, Kenneth C. Litkowski, and Suzanne Stevenson. 2008. Special issue introduction: Semantic role labeling: An introduction to the special issue. *Computational Linguistics*, 34(2), 145–159. https://doi.org/10.1162/coli.2008.34.2.145

Louis Martin, Éric de la Clergerie, Benoît Sagot, and Antoine Bordes. 2020. Controllable sentence simplification. In *Proceedings of The 12th Language Resources and Evaluation Conference*, pages 4689–4698, Marseille, France. European Language Resources Association. https://www.aclweb.org/anthology/2020.lrec-1.577

A.V. Martinet and A.J. Thomson. 1996. *A Practical English Grammar. 4th edition*. Oxford University Press.

Mausam, Michael Schmitz, Stephen Soderland, Robert Bart, and Oren Etzioni. 2012.Open language learning for information extraction. In *Proceedings of the 2012 Joint Conference on Empirical Methods in Natural Language Processing and Computational Natural Language Learning*, pages 523–534, Jeju Island, Korea. Association for Computational Linguistics. https://www.aclweb.org/anthology/D12-1048

Mausam Mausam. 2016. Open information extraction systems and downstream applications. In *Proceedings of the Twenty-Fifth International Joint Conference on Artificial Intelligence*, IJCAI'16, page 4074–4077. AAAI Press.

Filipe Mesquita, Jordan Schmidek, and Denilson Barbosa. 2013. Effectiveness and efficiency of open relation extraction. In *Proceedings of the 2013 Conference on Empirical Methods in Natural Language Processing*, pages 447–457, Seattle, Washington, USA. Association for Computational Linguistics. https://www.aclweb.org/anthology/D13-1043

Julian Michael, Gabriel Stanovsky, Luheng He, Ido Dagan, and Luke Zettlemoyer. 2018. Crowdsourcing question-answer meaning representations. In *Proceedings of the 2018 Conference of the North American Chapter of the Association for Computational Linguistics: Human Language Technologies, Volume 2 (Short Papers)*, pages 560–568, New Orleans, Louisiana. Association for Computational Linguistics. https://doi.org/10.18653/v1/N18-2089

Eleni Miltsakaki, Rashmi Prasad, Aravind Joshi, and Bonnie Webber. 2004. The Penn discourse treebank. In *Proceedings of the Fourth International Conference on Language Resources and Evaluation (LREC'04)*, Lisbon, Portugal. European Language Resources Association (ELRA). http://www.lrec-conf.org/proceedings/lrec2004/pdf/618.pdf

Ruslan Mitkov and Horacio Saggion. 2018. Text simplification. https://www.oxfordhandbooks.com/view/10.1093/oxfordhb/9780199573691.001.0001/oxfordhb-9780199573691-e-52

Makoto Miwa, Rune Sætre, Yusuke Miyao, and Jun'ichi Tsujii. 2010. Entity-focused sentence simplification for relation extraction. In *Proceedings of the 23rd International Conference on Computational Linguistics (Coling 2010)*, pages 788–796, Beijing, China. Coling 2010 Organizing Committee. https://www.aclweb.org/anthology/C10-1089

Tsendsuren Munkhdalai and Hong Yu. 2017. Neural semantic encoders. In *Proceedings of the 15th Conference of the European Chapter of the Association for Computational Linguistics: Volume 1, Long Papers*, pages 397–407, Valencia, Spain. Association for Computational Linguistics. https://www.aclweb.org/anthology/E17-1038

Shashi Narayan and Claire Gardent. 2014. Hybrid simplification using deep semantics and machine translation. In *Proceedings of the 52nd Annual Meeting of the Association for Computational Linguistics (Volume 1: Long Papers)*, pages 435–445, Baltimore, Maryland. Association for Computational Linguistics. https://doi.org/10.3115/v1/P14-1041

Shashi Narayan and Claire Gardent. 2016. Unsupervised sentence simplification using deep semantics. In *Proceedings of the 9th International Natural Language Generation conference*, pages 111–120, Edinburgh, UK. Association for Computational Linguistics. https://doi.org/10.18653/v1/W16-6620

Shashi Narayan, Claire Gardent, Shay B. Cohen, and Anastasia Shimorina. 2017. Split and rephrase. In *Proceedings of the 2017 Conference on Empirical Methods in Natural Language Processing*, pages 606–616, Copenhagen, Denmark. Association for Computational Linguistics. https://doi.org/10.18653/v1/D17-1064

Christina Niklaus, Bernhard Bermeitinger, Siegfried Handschuh, and André Freitas. 2016. A sentence simplification system for improving relation extraction. In *Proceedings of COLING 2016, the 26th International Conference on Computational Linguistics: System Demonstrations*, pages 170–174, Osaka, Japan. The COLING 2016 Organizing Committee. https://www.aclweb.org/anthology/C16-2036

Christina Niklaus, Matthias Cetto, André Freitas, and Siegfried Handschuh. 2018. A survey on open information extraction. In *Proceedings of the 27th International Conference on Computational Linguistics*, pages 3866–3878, Santa Fe, New Mexico, USA. Association for Computational Linguistics. https://www.aclweb.org/anthology/C18-1326

Christina Niklaus, Matthias Cetto, André Freitas, and Siegfried Handschuh. 2019a. DisSim: A discourse-aware syntactic text simplification framework for English and German. In *Proceedings of the 12th International Conference on Natural Language Generation*, pages 504–507, Tokyo, Japan. Association for Computational Linguistics. https://doi.org/10.18653/v1/W19-8662

Christina Niklaus, Matthias Cetto, André Freitas, and Siegfried Handschuh. 2019b. Transforming complex sentences into a semantic hierarchy. In *Proceedings of the 57th Annual Meeting of the Association for Computational Linguistics*, pages 3415–3427, Florence, Italy. Association for Computational Linguistics. https://doi.org/10.18653/v1/P19-1333

Christina Niklaus, André Freitas, and Siegfried Handschuh. 2019c. MinWikiSplit: A sentence splitting corpus with minimal propositions. In *Proceedings of the 12th International Conference on Natural Language Generation*, pages 118–123, Tokyo, Japan. Association for Computational Linguistics. https://doi.org/10.18653/v1/W19-8615

Sergiu Nisioi, Sanja Štajner, Simone Paolo Ponzetto, and Liviu P. Dinu. 2017. Exploring neural text simplification models. In *Proceedings of the 55th Annual Meeting of the Association for Computational Linguistics (Volume 2: Short Papers)*, pages 85–91, Vancouver, Canada. Association for Computational Linguistics. https://doi.org/10.18653/v1/P17-2014

Franz Josef Och and Hermann Ney. 2004. The alignment template approach to statistical machine translation. *Computational Linguistics*, 30(4):417–449. https://doi.org/10.1162/0891201042544884

Gustavo Paetzold, Fernando Alva-Manchego, and Lucia Specia. 2017. MASSAlign: Alignment and annotation of comparable documents. In *Proceedings of the IJCNLP 2017, System Demonstrations*, pages 1–4, Tapei, Taiwan. Association for Computational Linguistics. https://www.aclweb.org/anthology/I17-3001

Gustavo Paetzold and Lucia Specia. 2017. Lexical simplification with neural ranking. In *Proceedings of the 15th Conference of the European Chapter of the Association for Computational Linguistics: Volume 2, Short Papers*, pages 34–40, Valencia, Spain. Association for Computational Linguistics. https://www.aclweb.org/anthology/E17-2006

Gustavo H. Paetzold and Lucia Specia. 2013. Text simplification as tree transduction. In *Proceedings of the 9th Brazilian Symposium in Information and Human Language Technology*. https://www.aclweb.org/anthology/W13-4813

Gustavo H. Paetzold and Lucia Specia. 2016. Unsupervised lexical simplification for nonnative speakers. In *Proceedings of the Thirtieth AAAI Conference on Artificial Intelligence*, AAAI'16, page 3761–3767. AAAI Press.

Harinder Pal and Mausam. 2016. Demonyms and compound relational nouns in nominal open IE. In *Proceedings of the 5th Workshop on Automated Knowledge Base Construction*, pages 35–39, San Diego, CA. Association for Computational Linguistics. https://doi.org/10.18653/v1/W16-1307

Martha Palmer, Daniel Gildea, and Paul Kingsbury. 2005. The Proposition Bank: An annotated corpus of semantic roles. *Computational Linguistics*, 31(1), 71–106. https://doi.org/10.1162/0891201053630264

Kishore Papineni, Salim Roukos, Todd Ward, and Wei-Jing Zhu. 2002. Bleu: a method for automatic evaluation of machine translation. In *Proceedings of the 40th Annual Meeting of the Association for Computational Linguistics*, pages 311–318, Philadelphia, Pennsylvania, USA. Association for Computational Linguistics. https://doi.org/10.3115/1073083. 1073135

Ellie Pavlick and Chris Callison-Burch. 2016. Simple PPDB: A paraphrase database for simplification. In *Proceedings of the 54th Annual Meeting of the Association for Computational Linguistics (Volume 2: Short Papers)*, pages 143–148, Berlin, Germany. Association for Computational Linguistics. https://doi.org/10.18653/v1/P16-2024

Matt Post, Juri Ganitkevitch, Luke Orland, Jonathan Weese, Yuan Cao, and Chris Callison-Burch. 2013. Joshua 5.0: Sparser, better, faster, server. In *Proceedings of the Eighth Workshop on Statistical Machine Translation*, pages 206–212, Sofia, Bulgaria. Association for Computational Linguistics. https://www.aclweb.org/anthology/W13-2226

Rashmi Prasad, Eleni Miltsakaki, Nikhil Dinesh, Alan Lee, Aravind Joshi, Livio Robaldo, and Bonnie L Webber. 2007. The penn discourse treebank 2.0 annotation manual. Technical report, The PDTB Research Group.

Randolph Quirk, Sidney Greenbaum, Geoffrey Leech, and Jan Svartvik. 1985. *A Comprehensive Grammar of the English Language*. Longman, London.

Lance Ramshaw and Mitch Marcus. 1995. Text chunking using transformation-based learning. In *Third Workshop on Very Large Corpora*. https://www.aclweb.org/anthology/W95-0107

Scott Reed and Nando de Freitas. 2016. Neural programmer-interpreters. In *International Conference on Learning Representations (ICLR)*. http://arxiv.org/pdf/1511.06279v3

Ellen Riloff and Rosie Jones. 1999. Learning dictionaries for information extraction by multi-level bootstrapping. In *Proceedings of the Sixteenth National Conference on Artificial Intelligence and the Eleventh Innovative Applications of Artificial Intelligence Conference Innovative Applications of Artificial Intelligence*, AAAI '99/IAAI '99, page 474–479, USA. American Association for Artificial Intelligence.

Horacio Saggion, Sanja Štajner, Stefan Bott, Simon Mille, Luz Rello, and Biljana Drndarevic. 2015. Making it simplext: Implementation and evaluation of a text simplification system for spanish. *ACM Transactions on Accessible Computing*, 6(4). https://doi.org/10.1145/2738046

Swarnadeep Saha and Mausam. 2018. Open information extraction from conjunctive sentences. In *Proceedings of the 27th International Conference on Computational Linguistics*, pages 2288–2299, Santa Fe, New Mexico, USA. Association for Computational Linguistics. https://www.aclweb.org/anthology/C18-1194

Swarnadeep Saha, Harinder Pal, and Mausam. 2017. Bootstrapping for numerical open IE. In *Proceedings of the 55th Annual Meeting of the Association for Computational Linguistics (Volume 2: Short Papers)*, pages 317–323, Vancouver, Canada. Association for Computational Linguistics. https://doi.org/10.18653/v1/P17-2050

Evan Sandhaus. 2008. The New York Times Annotated Corpus. *Linguistic Data Consortium, Philadelphia*, 6(12).

Rob A. Van der Sandt. 1992. Presupposition Projection as Anaphora Resolution. *Journal of Semantics*, 9(4), 333–377. https://doi.org/10.1093/jos/9.4.333

Carolina Scarton, Gustavo Paetzold, and Lucia Specia. 2018. Text simplification from professionally produced corpora. In *Proceedings of the Eleventh International Conference on*

Language Resources and Evaluation (LREC 2018), Miyazaki, Japan. European Language Resources Association (ELRA). https://www.aclweb.org/anthology/L18-1553

Carolina Scarton and Lucia Specia. 2018. Learning simplifications for specific target audiences. In *Proceedings of the 56th Annual Meeting of the Association for Computational Linguistics (Volume 2: Short Papers)*, pages 712–718, Melbourne, Australia. Association for Computational Linguistics. https://doi.org/10.18653/v1/P18-2113

Jordan Schmidek and Denilson Barbosa. 2014. Improving open relation extraction via sentence re-structuring. In *Proceedings of the Ninth International Conference on Language Resources and Evaluation (LREC'14)*, pages 3720–3723, Reykjavik, Iceland. European Language Resources Association (ELRA). http://www.lrec-conf.org/proceedings/lrec2014/pdf/1038_Paper.pdf

Rudolf Schneider, Tom Oberhauser, Tobias Klatt, Felix A. Gers, and Alexander Löser. 2017. Analysing errors of open information extraction systems. In *Proceedings of the First Workshop on Building Linguistically Generalizable NLP Systems*, pages 11–18, Copenhagen, Denmark. Association for Computational Linguistics. https://doi.org/10.18653/v1/W17-5402

Natalie M. Schrimpf. 2018. Using rhetorical topics for automatic summarization. In *Proceedings of the Society for Computation in Linguistics (SCiL) 2018*, pages 125–135. https://doi.org/10.7275/R5SQ8XM6

Abigail See, Peter J. Liu, and Christopher D. Manning. 2017. Get to the point: Summarization with pointer-generator networks. In *Proceedings of the 55th Annual Meeting of the Association for Computational Linguistics (Volume 1: Long Papers)*, pages 1073–1083, Vancouver, Canada. Association for Computational Linguistics. https://doi.org/10.18653/v1/P17-1099

Rico Sennrich, Orhan Firat, Kyunghyun Cho, Alexandra Birch, Barry Haddow, Julian Hitschler, Marcin Junczys-Dowmunt, Samuel Läubli, Antonio Valerio Miceli Barone, Jozef Mokry, and Maria Nădejde. 2017. Nematus: a toolkit for neural machine translation. In *Proceedings of the Software Demonstrations of the 15th Conference of the European Chapter of the Association for Computational Linguistics*, pages 65–68, Valencia, Spain. Association for Computational Linguistics. https://www.aclweb.org/anthology/E17-3017

Matthew Shardlow. 2014. A survey of automated text simplification. *International Journal of Advanced Computer Science and Applications*, 4(1):58–70.

Advaith Siddharthan. 2002. An architecture for a text simplification system. In *Language Engineering Conference, 2002. Proceedings*, pages 64–71. IEEE.

Advaith Siddharthan. 2006. Syntactic simplification and text cohesion. *Research on Language and Computation*, 4(1):77–109.

Advaith Siddharthan. 2014. A survey of research on text simplification. *ITL-International Journal of Applied Linguistics*, 165(2):259–298.

Advaith Siddharthan and Angrosh Mandya. 2014. Hybrid text simplification using synchronous dependency grammars with hand-written and automatically harvested rules. In *Proceedings of the 14th Conference of the European Chapter of the Association for Computational Linguistics*, pages 722–731, Gothenburg, Sweden. Association for Computational Linguistics. https://doi.org/10.3115/v1/E14-1076

Advaith Siddharthan, Ani Nenkova, and Kathleen McKeown. 2004. Syntactic simplification for improving content selection in multi-document summarization. In *COLING 2004:*

Proceedings of the 20th International Conference on Computational Linguistics, pages 896–902, Geneva, Switzerland. COLING. https://www.aclweb.org/anthology/C04-1129

David Smith and Jason Eisner. 2006. Quasi-synchronous grammars: Alignment by soft projection of syntactic dependencies. In *Proceedings on the Workshop on Statistical Machine Translation*, pages 23–30, New York City. Association for Computational Linguistics. https://www.aclweb.org/anthology/W06-3104

Richard Socher, John Bauer, Christopher D. Manning, and Andrew Y. Ng. 2013. Parsing with compositional vector grammars. In *Proceedings of the 51st Annual Meeting of the Association for Computational Linguistics (Volume 1: Long Papers)*, pages 455–465, Sofia, Bulgaria. Association for Computational Linguistics. https://www.aclweb.org/anthology/P13-1045

Yangqiu Song and Dan Roth. 2015. Unsupervised sparse vector densification for short text similarity. In *Proceedings of the 2015 Conference of the North American Chapter of the Association for Computational Linguistics: Human Language Technologies*, pages 1275–1280, Denver, Colorado. Association for Computational Linguistics. https://doi.org/10.3115/v1/N15-1138

Radu Soricut and Daniel Marcu. 2003. Sentence level discourse parsing using syntactic and lexical information. In *Proceedings of the 2003 Human Language Technology Conference of the North American Chapter of the Association for Computational Linguistics*, pages 228–235. https://www.aclweb.org/anthology/N03-1030

Lucia Specia. 2010. Translating from complex to simplified sentences. In *Proceedings of the 9th International Conference on Computational Processing of the Portuguese Language*, PROPOR'10, pages 30–39, Berlin, Heidelberg. Springer-Verlag.

Sanja Štajner and Goran Glavaš. 2017. Leveraging event-based semantics for automated text simplification. *Expert systems with applications*, 82:383–395.

Sanja Štajner and Sergiu Nisioi. 2018. A detailed evaluation of neural sequence-to-sequence models for in-domain and cross-domain text simplification. In *Proceedings of the Eleventh International Conference on Language Resources and Evaluation (LREC 2018)*, Miyazaki, Japan. European Language Resources Association (ELRA). https://www.aclweb.org/anthology/L18-1479

Sanja Štajner and Maja Popovic. 2016. Can text simplification help machine translation? In *Proceedings of the 19th Annual Conference of the European Association for Machine Translation*, pages 230–242. https://www.aclweb.org/anthology/W16-3411

Sanja Štajner and Maja Popović. 2018. Improving machine translation of English relative clauses with automatic text simplification. In *Proceedings of the 1st Workshop on Automatic Text Adaptation (ATA)*, pages 39–48, Tilburg, the Netherlands. Association for Computational Linguistics. https://doi.org/10.18653/v1/W18-7006

Gabriel Stanovsky and Ido Dagan. 2016. Creating a large benchmark for open information extraction. In *Proceedings of the 2016 Conference on Empirical Methods in Natural Language Processing*, pages 2300–2305, Austin, Texas. Association for Computational Linguistics. https://doi.org/10.18653/v1/D16-1252

Gabriel Stanovsky, Jessica Ficler, Ido Dagan, and Yoav Goldberg. 2016. Getting more out of syntax with props. *CoRR*, abs/1603.01648. http://arxiv.org/abs/1603.01648

Gabriel Stanovsky, Julian Michael, Luke Zettlemoyer, and Ido Dagan. 2018. Supervised open information extraction. In *Proceedings of the 2018 Conference of the North American Chapter of the Association for Computational Linguistics: Human Language Technolo-*

gies, Volume 1 (Long Papers), pages 885–895, New Orleans, Louisiana. Association for Computational Linguistics. https://doi.org/10.18653/v1/N18-1081

Elior Sulem, Omri Abend, and Ari Rappoport. 2018a. BLEU is not suitable for the evaluation of text simplification. In *Proceedings of the 2018 Conference on Empirical Methods in Natural Language Processing*, pages 738–744, Brussels, Belgium. Association for Computational Linguistics. https://doi.org/10.18653/v1/D18-1081

Elior Sulem, Omri Abend, and Ari Rappoport. 2018b. Semantic structural evaluation for text simplification. In *Proceedings of the 2018 Conference of the North American Chapter of the Association for Computational Linguistics: Human Language Technologies, Volume 1 (Long Papers)*, pages 685–696, New Orleans, Louisiana. Association for Computational Linguistics. https://doi.org/10.18653/v1/N18-1063

Elior Sulem, Omri Abend, and Ari Rappoport. 2018c. Simple and effective text simplification using semantic and neural methods. In *Proceedings of the 56th Annual Meeting of the Association for Computational Linguistics (Volume 1: Long Papers)*, pages 162–173, Melbourne, Australia. Association for Computational Linguistics. https://doi.org/10.18653/v1/P18-1016

Mihai Surdeanu. 2013. Overview of the tac2013 knowledge base population evaluation: English slot filling and temporal slot filling. *Theory and Applications of Categories*.

Sai Surya, Abhijit Mishra, Anirban Laha, Parag Jain, and Karthik Sankaranarayanan. 2019. Unsupervised neural text simplification. In *Proceedings of the 57th Annual Meeting of the Association for Computational Linguistics*, pages 2058–2068, Florence, Italy. Association for Computational Linguistics. https://doi.org/10.18653/v1/P19-1198

Julia Suter, Sarah Ebling, and Martin Volk. 2016. Rule-based automatic text simplification for german. In *13th Conference on Natural Language Processing (KONVENS 2016)*. University of Zurich. https://doi.org/10.5167/uzh-128601

Ilya Sutskever, Oriol Vinyals, and Quoc V. Le. 2014. Sequence to sequence learning with neural networks. In *Proceedings of the 27th International Conference on Neural Information Processing Systems—Volume 2*, NIPS'14, page 3104–3112, Cambridge, MA, USA. MIT Press.

Maite Taboada and Debopam Das. 2013. Annotation upon annotation: Adding signalling information to a corpus of discourse relations. *D&D*, 4(2):249–281.

Maite Taboada and William C. Mann. 2006. Rhetorical structure theory: looking back and moving ahead. *Discourse Studies*, 8(3):423–459. https://doi.org/10.1177/1461445606061881

Johan Van Benthem. 2008. *A brief history of natural logic*. LondonCollege Publications.

Ashish Vaswani, Noam Shazeer, Niki Parmar, Jakob Uszkoreit, Llion Jones, Aidan N. Gomez, undefinedukasz Kaiser, and Illia Polosukhin. 2017. Attention is all you need. In *Proceedings of the 31st International Conference on Neural Information Processing Systems*, NIPS'17, page 6000–6010, Red Hook, NY, USA. Curran Associates Inc.

David Vickrey and Daphne Koller. 2008. Sentence simplification for semantic role labeling. In *Proceedings of ACL-08: HLT*, pages 344–352, Columbus, Ohio. Association for Computational Linguistics. https://www.aclweb.org/anthology/P08-1040

Tu Vu, Baotian Hu, Tsendsuren Munkhdalai, and Hong Yu. 2018. Sentence simplification with memory-augmented neural networks. In *Proceedings of the 2018 Conference of the North American Chapter of the Association for Computational Linguistics: Human Language*

Technologies, Volume 2 (Short Papers), pages 79–85, New Orleans, Louisiana. Association for Computational Linguistics. https://doi.org/10.18653/v1/N18-2013

Yizhong Wang, Sujian Li, and Houfeng Wang. 2017. A two-stage parsing method for text-level discourse analysis. In *Proceedings of the 55th Annual Meeting of the Association for Computational Linguistics (Volume 2: Short Papers)*, pages 184–188, Vancouver, Canada. Association for Computational Linguistics. https://doi.org/10.18653/v1/P17-2029

Aaron Steven White, Drew Reisinger, Keisuke Sakaguchi, Tim Vieira, Sheng Zhang, Rachel Rudinger, Kyle Rawlins, and Benjamin Van Durme. 2016. Universal decompositional semantics on universal dependencies. In *Proceedings of the 2016 Conference on Empirical Methods in Natural Language Processing*, pages 1713–1723, Austin, Texas. Association for Computational Linguistics. https://doi.org/10.18653/v1/D16-1177

Kristian Woodsend and Mirella Lapata. 2011. Learning to simplify sentences with quasi-synchronous grammar and integer programming. In *Proceedings of the 2011 Conference on Empirical Methods in Natural Language Processing*, pages 409–420, Edinburgh, Scotland, UK. Association for Computational Linguistics. https://www.aclweb.org/anthology/D11-1038

Fei Wu and Daniel S. Weld. 2010. Open information extraction using Wikipedia. In *Proceedings of the 48th Annual Meeting of the Association for Computational Linguistics*, pages 118–127, Uppsala, Sweden. Association for Computational Linguistics. https://www.aclweb.org/anthology/P10-1013

Sander Wubben, Antal van den Bosch, and Emiel Krahmer. 2012. Sentence simplification by monolingual machine translation. In *Proceedings of the 50th Annual Meeting of the Association for Computational Linguistics (Volume 1: Long Papers)*, pages 1015–1024, Jeju Island, Korea. Association for Computational Linguistics. https://www.aclweb.org/anthology/P12-1107

Wei Xu, Chris Callison-Burch, and Courtney Napoles. 2015. Problems in current text simplification research: New data can help. *Transactions of the Association for Computational Linguistics*, 3:283–297. https://doi.org/10.1162/tacl_a_00139

Wei Xu, Courtney Napoles, Ellie Pavlick, Quanze Chen, and Chris Callison-Burch. 2016. Optimizing statistical machine translation for text simplification. *Transactions of the Association for Computational Linguistics*, 4:401–415. https://doi.org/10.1162/tacl_a_00107

Ying Xu, Mi-Young Kim, Kevin Quinn, Randy Goebel, and Denilson Barbosa. 2013. Open information extraction with tree kernels. In *Proceedings of the 2013 Conference of the North American Chapter of the Association for Computational Linguistics: Human Language Technologies*, pages 868–877, Atlanta, Georgia. Association for Computational Linguistics. https://www.aclweb.org/anthology/N13-1107

Kenji Yamada and Kevin Knight. 2001. A syntax-based statistical translation model. In *Proceedings of the 39th Annual Meeting of the Association for Computational Linguistics*, pages 523–530, Toulouse, France. Association for Computational Linguistics. https://doi.org/10.3115/1073012.1073079

Taha Yasseri, Andras Kornai, and Janos Kertesz. 2012. A practical approach to language complexity: a wikipedia case study. *PloS one*, 7(11).

Xingxing Zhang and Mirella Lapata. 2017. Sentence simplification with deep reinforcement learning. In *Proceedings of the 2017 Conference on Empirical Methods in Natural Language Processing*, pages 584–594, Copenhagen, Denmark. Association for Computational Linguistics. https://doi.org/10.18653/v1/D17-1062

Sanqiang Zhao, Rui Meng, Daqing He, Andi Saptono, and Bambang Parmanto. 2018. Integrating transformer and paraphrase rules for sentence simplification. In *Proceedings of the 2018 Conference on Empirical Methods in Natural Language Processing*, pages 3164–3173, Brussels, Belgium. Association for Computational Linguistics. https://doi.org/10.18653/v1/D18-1355

Zhemin Zhu, Delphine Bernhard, and Iryna Gurevych. 2010. A monolingual tree-based translation model for sentence simplification. In *Proceedings of the 23rd International Conference on Computational Linguistics (Coling 2010)*, pages 1353–1361, Beijing, China. Coling 2010 Organizing Committee. https://www.aclweb.org/anthology/C10-1152